Many good books have been written on healthy building, but until now there has not been a nitty-gritty reference manual that covers everything from theory to specification language in a way that can be applied to any construction type. *Prescriptions for a Healthy House* is introductory enough to be used by someone new to the field, yet detailed and practical enough to be a valuable reference for the more experienced.

The best thing about this book is that it is laid out to be used. The graphic design allows for easy perusal to find the charts, case studies, specification language, resources, details, or supporting text. In fact, I made use of *Prescriptions* the first day I got it: a client called with a question, and I turned straight to the relevant page and read her a concise list of practical suggestions. I was relieved not to have to comb my mental or physical database!

The backgrounds of the authors—experienced architect, M.D., and healthy building consultant—combine to give the book a breadth and depth rarely found in one place. More than an admonition to go nontoxic or a list of materials, the book includes practical strategies and procedures, clearly gained from experience, to ensure that the finished home is a haven, not a nightmare. The case studies bring home the authors' points. When you read that a cleanup product caused a nearly finished house to be uninhabitable, you know *why* you need to specify everything that is used on the site. In fact, it makes me want to specify that every contractor read this book!

—Carol Venolia, publisher of *Building with Nature* newsletter
and author of *Healing Environments*

I liked this book very much—it fills a real need.

—Vincent A. Marinkovich, M.D.
Immunologist, allergist, and pediatrician,
Menlo Park, CA

I will definitely recommend this book to my patients.

—William Shrader, M.D.
Allergist, specialist in environmental medicine,
Santa Fe, NM

Prescriptions for a Healthy House:

A Practical Guide for Architects, Builders and Homeowners

Paula Baker, A.I.A.,
Erica Elliott, M.D.,
and John Banta, B.A.

Illustrations by Lisa Flynn

INWORD PRESS

Prescriptions for a Healthy House:
A Practical Guide for Architects, Builders and Homeowners

By Paula Baker, Erica Elliott, and John Banta
Illustrated by Lisa Flynn

Published by:
InWord Press
2530 Camino Entrada
Santa Fe, NM 87505-4835 USA

This book was printed using recycled, acid-free paper and soy-based inks.

First Edition, 1998
SAN 694-0269
10 9 8 7 6 5 4 3 2 1
Printed in the United States of America

Library of Congress Cataloging-in-Publication Data

Baker, Paula, 1953-
Prescriptions for a healthy house : a practical guide for architects, builders, and homeowners / by Paula Baker, Erica Elliott, and John Banta.
p. cm.
Includes bibliographical references and index.
ISBN 1-56690-355-6
1. House construction. 2. Sick building syndrome. 3. Green products.
I. Elliott Erica, 1948- II. Banta, John, 1955- III. Title.
TH4812.B35 1997
696–dc21

97-42428
CIP

Contents

Division 3 - Concrete 61

Division 4 - Masonry or Other Alternatives to Frame Construction 67

Division 5 - Metals 75

Division 6 - Wood and Plastics 77

Division 7 - Thermal and Moisture Control 93

Division 8 - Openings 111

Division 9 - Finishes 117

Warning and Disclaimer

InWord Press is a registered trademark of High Mountain Press, Inc. Many terms are mentioned in this book believed to be trademarks or service marks have been appropriately capitalized. InWord Press cannot attest to the accuracy of this information. Use of a term in this book should not be regarded as affecting the validity of any trademark or service mark. InWord Press and the authors make no claim to these marks.

If you have multiple chemical sensitivities (MCS) or environmental illness (see Appendix 1 for definitions), then it is especially important to make the one environment that you can control, your home, as free from toxic substances as possible. This book will provide some general guidelines of which you should be aware. Please do not assume that the materials or products listed in this book will be appropriate for your particular sensitivities. Construction materials suggested in this book that are generally considered to be safer could nevertheless elicit symptoms in you. It is important that you test each substance that you will use in a home for your level of reaction. If you are embarking on a construction project, we advise you to work with an architect and physician who understand MCS.

About the Authors

Paula Baker is the primary author of this book. As an architect, Paula is intimately familiar with the materials and methods of standard construction. As a baubiologist, she also knows where these practices are in conflict with human health and which alternatives are available. Having designed and supervised the construction of healthy homes, she is well-versed in the challenges presented when one deviates from accepted construction norms. It was her vision to bring diverse information together into a practical reference book. Her collaboration with Erica and John has enabled this vision to become a reality.

As a physician trained in both family practice and environmental medicine, **Erica Elliott** has extensive clinical experience in the medical consequences of standard construction. In this book she has interpreted the world of medicine and chemistry so that the reader can begin to understand the complex relationship between chemical exposure in the indoor environment and human health. As a talented linguist, she has tirelessly lent her skills to this book by editing and adding clarity to a rather complicated subject matter.

John Banta brings invaluable insight to the topics covered in this book, gained in the course of over a decade of experience in troubleshooting indoor environmental problems. His expertise covers many aspects of indoor air quality, including the detection and reduction of electromagnetic fields, and the recognition and abatement of mold problems.

John holds a degree in environmental science.

Acknowledgments

The authors wish to thank the many people who have offered their guidance, expertise, and encouragement in the completion of this book. Special thanks go to Pauline Kenny for her tireless efforts and computer wizardry which helped to transform the data into something that resembled a book. Our gratitude goes to Will and Louise Pape who graciously offered their ranch as a working retreat center and offered practical advice and inspiration each step of the way. Thanks are due to Santa Fe consultants Greg Friedman of the Good Water Company and Carl Rosenberg of Sunspot Design who provided useful information about water filtration and mechanical systems, respectively. And thanks to members of the Healthy Housing Coalition, along with friends, family, patients, and clients who prodded us along with the recurrent refrain, "Is it finished yet?"

And finally, we wish to offer special acknowledgment to Helmut Zeihe, founder of the Institute for Baubiology and Ecology in Clearwater, Florida, teacher and mentor of many of us who are concerned about healthy homes.

—July 1997
Paula Baker, A.I.A.,
Erica Elliott, M.D.,
& John Banta, B.A.

InWord Press...

Dan Raker, President
Dale Bennie, Vice President
Scott Brassart, Acquisitions Editor
Carol Leyba, Associate Publisher
Barbara Kohl, Acquisitions Editor
Cynthia Welch, Production Manager
Laurel Avery, Production Editor
Liz Bennie, Marketing Director
Lauri Hogan, Marketing Services Manager
Lynne Egensteiner, Cover designer, Illustrator

Foreword

It has been said that we shape our buildings, and then our buildings shape us. Upon considering the fact that the average American spends 90% or more of life indoors, the significance of this statement becomes apparent. In this era of unprecedented technological advancement, it stands to reason that we would use our knowledge to create indoor environments with exceptional vitality which could enhance our health and sense of well-being. Yet this has not been the case. The U.S. Environmental Protection Agency (EPA) has recently stated that "indoor air pollution in residences, offices, schools, and other buildings is widely recognized as one of the most serious potential environmental risks to human health" and is, in fact, many times more of a health threat than outdoor air pollution.

How has this sad state of affairs developed? Since the oil embargo of 1973, we have placed a high priority on energy efficiency, creating buildings that are increasingly airtight. Concurrently, the building industry has introduced inexpensive synthetic building products to the public and furnishings that are mass-produced and require little maintenance. Until very recently, little attention has been paid to the toxicity of these products, allowing consumers to remain largely ignorant of the health threat that they pose.

The average person has little background in chemistry and a false assumption that, in order for building products to be allowed on the market, they must be reasonably safe. The disturbing truth is that, according to the EPA, of the more than 80,000 chemicals common in commercial use today, less than 1,000 have been tested for toxic effects on the human nervous system. The limited testing that has been implemented rarely takes into consideration the ongoing, low-level exposure to hundreds of chemicals we inhale or absorb simultaneously throughout our daily lives.

The toll on human health resulting from exposure to the chemical soup surrounding us is finally becoming clear. In 1986 The National Academy of Science estimated that 15% of the population suffered from chemical sensitivities. Based on current unofficial reports by physicians specializing in environmental medicine, that number is rising rapidly. These figures do not include people who unknowingly suffer from problems either directly or indirectly related to chronic, low-level toxic exposure. All too often symptoms are falsely attributed to the normal aging process.

Continued exposure to toxins in the indoor environment, even at low levels, has been linked to a vast spectrum of illnesses, ranging from chronic sinus infections, headaches, insomnia, anxiety, and joint pain, to full-blown multiple chemical sensitivities and other immune system disorders. In spite of overwhelming evidence of the health risks, the majority of new construction in the United States continues to create environments that harm human health.

There is, in fact, nothing complicated about erecting a healthy building. The solution is composed of many simple, but important steps. Many safer alternative materials and methods of design and building are becoming readily available. Nevertheless, the homeowner who desires to create a healthy building or remodel an existing one is still a pioneer facing three major obstacles.

1. Building for health is not the current standard of the construction industry. Although most architects and builders are now aware that health problems are associated with standard building practices, the industry in general has not responded with appropriate changes in these standards. There are no set and sanctioned prescriptions to follow for healthy building.

2. The homeowner receives false information. Most building professionals are uninformed about the details of healthful design and building. The prospective client who has heard about healthy building is often ill advised by professionals that there is either no need for concern or that healthy building is cost prohibitive.

3. There is a dearth of concise information. If homeowners are still committed to creating a healthy house and have managed to find an architect and builder who are receptive to working with them, then they must together undertake the daunting task of educating themselves. Distilling enough information to create a set of specifications for a project is an undertaking requiring extensive time and dedication.

The purpose of this book is to take the mystery out of healthy house building by walking the owner/architect/builder team through the construction process. It explains where and why standard building practices are not healthful, what to do differently, and how to obtain alternative materials and expertise.

The authors hope that you will find *Prescriptions for a Healthy House* a useful tool in your quest for healthier living.

This book is dedicated
to the millions of people
who are chronically ill
from chemical exposures.
May we be forewarned
and learn from your suffering.

Part I Overview

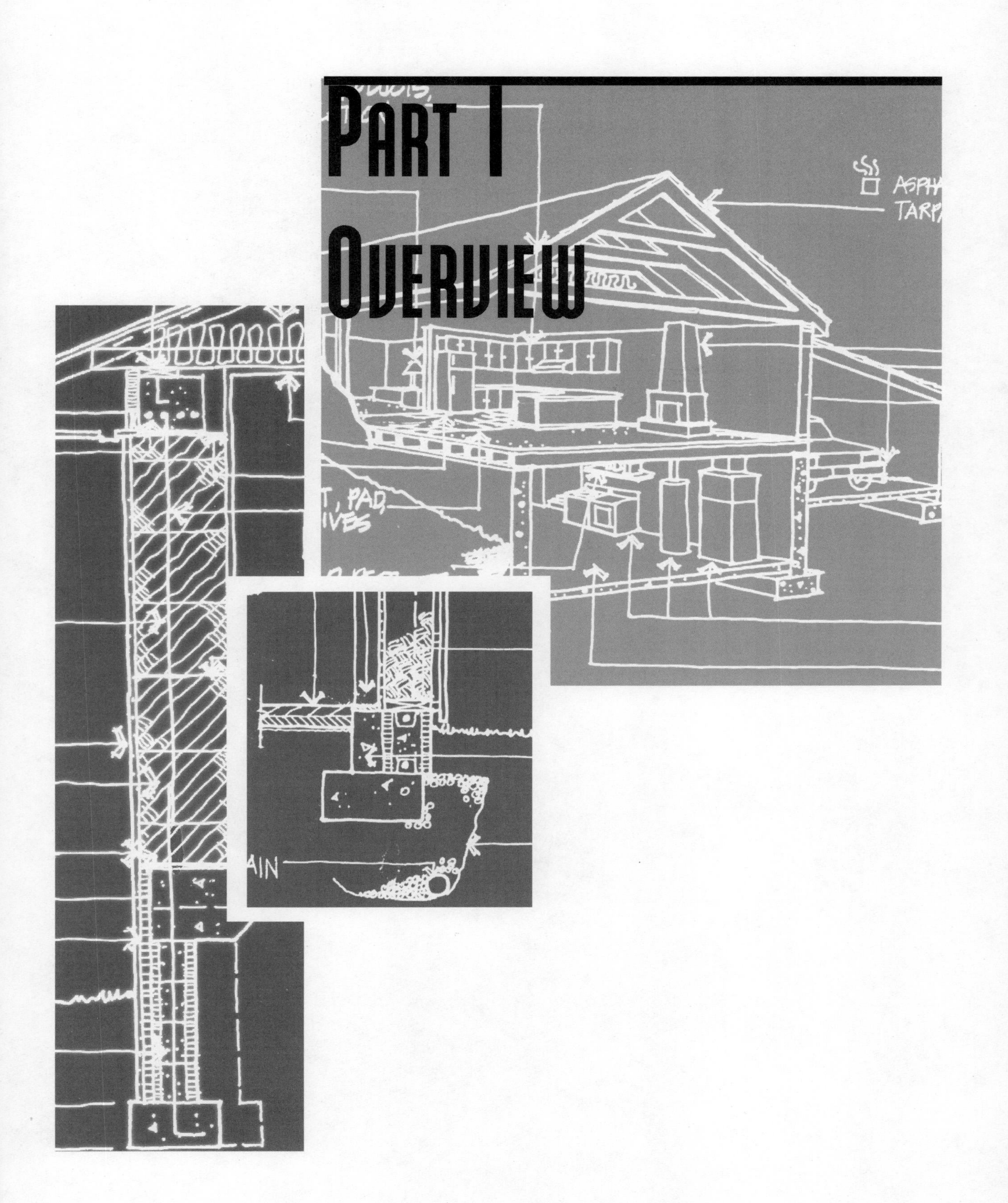

Description of the Problem and Solutions

Until about 25 years ago, indoor air pollution was a very limited phenomenon. Since that time, two basic things have changed in the way that buildings are constructed. First, thousands of chemicals have been incorporated into building materials. Second, buildings are sealed so tightly that the chemicals remain trapped inside homes, where the inhabitants inhale them into the lungs and absorb them into the skin. Prior to the energy crisis, the typical home averaged approximately one air exchange per hour. Now, in a well-sealed home, the air is often exchanged as little as once every five hours or longer, and that is not enough to ensure healthful air quality.

Two basic schools of thought have arisen as to how to solve the indoor pollution problem. The first approach involves eliminating as many pollutants as possible from within the building envelope and sealing it very tightly on the inside so that there is less need to worry about the chemical composition of the structure or insulation. Clean, filtered air is then mechanically pumped in, keeping the house under a slightly positive pressure so that air infiltration is controlled. Thus, the residents isolate themselves from a toxic world. If you do not have the luxury of clean, vital, and refreshing natural surroundings, then a certain amount of isolation and filtering may be essential.

The second school of thought involves building the structure out of natural or nontoxic materials that "breathe." Double adobe with natural insulation in the cavities, straw/clay, and straw bale are examples of building systems that fall into this category. The building is seen as a third skin (clothes being the second). As such, the building is viewed as a permeable organism interacting with the natural world and facilitating a balanced exchange of air and humidity. This approach is based on the precepts of *baubiology.*

On a philosophical level, we find this approach more appealing because it is based on an interactive relationship with the surrounding environment. In the context of global or regional ecology, we believe that our survival as a species is dependent on interaction and linking rather than isolationism.

Sealed construction is a relatively new concept, dating from the same time period as the invention of plastic. Contemporary building codes are based on sealed construction methods and results. In contrast, people have sheltered themselves in breathable structures throughout human history. Ironically, when you apply for a building permit, many breathing wall building systems will not be permitted, or if permitted, they are classified as experimental structures.

Whereas the first approach to reducing indoor pollution has the home isolated from a toxic world, in the second approach the home is interactive with the surrounding environment. One approach strives to create a healthy environment through technology, and the other through a return to nature. Much of the information in this book is applicable to either approach because the purpose is to identify and eliminate sources of indoor air pollution commonly found in the home. Although our emphasis will be on home construction, many of the same problems and solutions apply to schools and workplaces.

How Much More Will It Cost to Build a Healthy Home?

This is frequently the first question posed to Paula by her clients. The answer usually lies somewhere between zero and 25% more than standard construction.

Assume for a moment that you are house hunting. Your real estate agent contacts you and is very excited about having found a real bargain, a house going for 20% less than market value. Upon further inquiry, you learn that the house contains lead paint, asbestos insulation, and sits on a bed of radon emitting granite. If your reaction would be to snatch it up, then read no further because this book will be of little interest to you. This book is about how to avoid substances that are as harmful to your health as lead, asbestos, or radon, but which are commonly used in construction today.

In some cases, little or no extra money is required to use less toxic substances and methods to build and maintain a healthy home. A few examples are listed below.

- Additive free concrete costs no more than using concrete with toxic admixtures, provided that climatic conditions are appropriate for the project.
- Circuit breaker panels with reduced field configurations are the same price as other panels.

- Shortening wiring runs through careful planning will not only reduce exposure to electromagnetic fields, it will also save money.

- Unscented and nonchlorinated cleaning products cost no more and can be just as effective as compounds containing harsh chemicals.

In other cases such as the pair mentioned below, healthier alternatives are more expensive initially, but they are more economical in the long run.

- The most inexpensive type of roofing to install is comprised of tar and gravel, but its useful life is much shorter than many of the less toxic roofing systems specified in the discussion under "Division 7 - Thermal and Moisture Control."

- Although forced air is less expensive to install, a properly designed gas fired, hydronic radiant floor heating system is not only more comfortable and healthier, it is also virtually maintenance free. Higher initial installation costs will be outweighed over time by lower heating bills.

In some areas your decision to "go healthy" will cost more, and you will be faced with some difficult choices. We will try to offer you facts and alternatives so that your choices can be well-informed. There is no right answer in many instances. Sometimes your decision will come down to a trade-off between luxury and health. But then, what is luxury without health? You could ultimately spend a fortune on medical bills and lose quality of life, as have the people who have shared their stories with us. Furthermore, the cost to the environment of many current building practices is astronomical. Our children and grandchildren will ultimately pay the heaviest price.

Sources of Indoor Pollution

Indoor air pollutants can be classified into five main categories: volatile organic compounds (VOCs), toxic by-products of combustion, pesticides, electromagnetic field pollution, and naturally occurring pollutants. Each category of pollutant is described in the following sections.

Volatile Organic Compounds (VOCs)

Organic compounds are chemicals containing carbon hydrogen bonds at the molecular level. They are both naturally occurring and manufactured. Most synthetic organic compounds are petrochemicals, that is, derived from oil, gas, and coal.

Organic compounds can exist in the form of a gas, liquid, or solid particles. Substances that readily release vapors at room temperature are called volatile organic compounds. This outgassing is a form of evaporation of volatile compounds contained in solid material and results in a slow release of chemicals into the air.

At present, about 80,000 synthetic organic compounds are commercially available, and thousands more are produced annually by the chemical industry.[1] VOCs constitute a major source of toxic overload and can threaten individual health. Any organ of the body can be affected. Some of the more common symptoms include rashes, headaches, eye irritation, chronic cough, chronic sinus infections, joint and muscle pain, memory loss, inability to concentrate, irritability, fatigue, anxiety, depression, and increasing number of allergies.

Organic compounds can be classified into three categories based on derivation from petroleum products. The primary organic compounds include components directly derived from gas, oil, and coal and include propane, butane, benzene, xylene, paraffins, toluene, and styrene. These products are then used to derive the intermediate substances such as formaldehyde, phenols, acetone, isopropanol, and acetaldehyde. The end products produced from crude oil and natural gas include solvents, waxes, lacquers, synthetic detergents, synthetic fibers, and paints. Common sources of volatile organic compounds occurring in the indoor environment are listed below.

- plywood
- particle board
- wood paneling
- carpets and carpet pads
- insulation
- paints
- finishes
- solvents
- adhesives
- synthetic fabrics
- cleaning products
- body care products
- mothballs
- insecticides
- aerosol products
- art and hobby materials
- dry cleaned garments
- air fresheners

Interior view of kitchen shows deep window seat in bale walls, plaster interiors, and pine floors. All finishes are nontoxic. (Architect: Paula Baker. Contractor: Prull & Assoc. Inc. Photo: Julie Dean.)

Some of the more chemically sensitive people react to naturally occurring VOCs as well, such as terpenes which outgas from wood products. These individuals are urged to test their reactions to each product before making a major purchase, even if the product derives from a natural source.

You are undoubtedly familiar with the distinctive smell of a new house. The odor is composed primarily of outgassing chemicals from toxic volatile organic compounds. Some building products now report the parts per million of VOCs on labels, but this information can be misleading. Yes, it is true that the fewer parts per million the better, but certain chemicals such as dioxin are not safe in any detectable amount.[2] One of the goals in constructing a healthy house is to reduce the use of VOCs.

Timberframe and straw/clay home features natural mud plasters, and soilcrete floors with radiant floor heating. (Architect: Paula Baker. Timberframe and straw/clay: Robert Laporte. Rumford adobe fireplace and finishes: Charlie Carruthers. Photo: Julie Dean.)

Toxic By-products of Combustion

Gas, oil, coal, wood, and other fuels burned indoors consume valuable indoor oxygen unless air for combustion is supplied from the outdoors. All combustion appliances contribute to air pollution. In tight, energy efficient buildings, these fumes can cause serious health consequences.

Indoor combustion is found in fireplaces, wood stoves, gas-fired appliances such as ranges, clothes dryers, water heaters, furnaces, gas- and kerosene-fired space heaters, and oil and kerosene lamps. Some of the potentially harmful emissions include nitrogen dioxide, nitrous oxide, sulfur oxides, hydrogen cyanide, carbon monoxide, carbon dioxide, formaldehyde, particulate matter, and hydrocarbons from natural gas fumes such as butane, propane, pentane, methyl pentane, benzene, and xylene. The indoor levels of these gases are determined by the amount of fuel burned and the rate of exchange with outdoor air.

What are some of the potential health effects of combustion by-product gases? In a study of 47,000 chemically sensitive patients, the most important sources of indoor air pollution responsible for generating illness were the gas stove, hot water heater, and furnace.[3] Hazardous fumes can leak at the pipe joints and remain undetected, especially if they occur under flooring. In addition, every pilot light adds fumes, and the burning process itself releases fumes into the air. The primary effects of exposure to gas fumes are on the cardiovascular and nervous system, but can affect any organ of the body. Some of the earliest symptoms from exposure to gas fumes include depression, fatigue, irritability, and inability to concentrate.

Carbon monoxide is commonly produced during combustion, especially from gas fueled appliances. Carbon monoxide quickly diffuses throughout the entire house. Typically, these appliances must be removed from the homes of chemically sensitive patients to restore their health. Chronic exposure can result in multiple chemical sensitivities because carbon monoxide has the ability to interfere with the detoxification pathways in the liver, allowing the accumulation of toxic substances. Other effects of chronic carbon monoxide exposure include heart arrhythmia, decreased cognitive abilities, confusion, and fatigue.

Carbon dioxide is produced from burning natural gas. Elevated levels result in decreased mental acuity, loss of vigor, and fatigue. Nitrogen oxides are also released from gas appliances. A major source of contamination is the gas stove. Older models with pilot lights are particularly problematic. These gases are known to impact the nervous and reproductive systems.

Coal and wood burning fireplaces emit particulate matter as well as toxic fumes. They also consume indoor oxygen unless fresh outdoor air is supplied to them. Particles not expelled by blowing or sneezing can find their way into the lungs where they can remain for years.

It is important to mention that while parking or operating an automobile in an attached garage, gas, oil, and other volatile organic compounds diffuse into the structure and will affect indoor air quality in the home. Garages must be properly isolated from the main structure.

Pesticides

Although some pesticides may technically be considered VOCs, these often odorless and invisible substances have become such a health threat that they warrant a separate discussion. Pesticides, or biocides, are poisons designed to kill a variety of plants and animals such as insects (insecticides), weeds (herbicides), and mold or fungus (fungicides). They were first developed as offshoots of nerve gas used during World War II. Most pesticides are synthetic chemicals made from petroleum. They are composed of active ingredients, the chemical compounds designed to kill the target organism; and inert ingredients, chemicals that deliver the active ingredients to the target, preserve them, or make them easier to apply.

Many people assume that the pesticides they buy, or those used by lawn and pest control companies, are "safe." They assume the government is protecting them; that pesticides are scientifically tested; that if used according to the instructions on the label no harm will be done; and that the products would not be on the market if they were unsafe. All of these assumptions are incorrect.

EPA registration does not signify pesticide safety.[4] The EPA approves pesticides based on efficacy, not safety. Efficacy means the pesticide will kill the targeted pest. Out of the hundreds of active ingredients registered with the EPA, less than a dozen have been adequately tested for safety.[5] In fact, it is a violation of federal law to state or imply that the use of a pesticide is "safe when used as directed."

Inert ingredients, which can account for up to 99 percent of a pesticide, are not usually identified on the label. The Trade Secrets Act protects manufacturers from being required to fully disclose ingredients, even if the inert ingredients are potentially hazardous to human health. No studies of any kind are required on the inert ingredients. Many inert ingredients can be more toxic than the active ingredients, yet warning labels only apply to the active ingredients. In a Freedom of Information Act lawsuit, the Northwest Coalition for Alternatives to Pesticides (NCAP) obtained from the EPA a list of 1,400 of the 2,000 substances being used as inert ingredients in pesticides. These ingredients included Chicago sludge and other hazardous waste, asbestos, and some banned chemicals such as DDT.[6]

A recent study found that combining pesticides can make them up to 1,600 times more potent.[7] A good illustration of this synergy is found in a class of pesticides called pyrethroids which are mistakenly thought to be harmless because they are plant-derived. The unlabeled inert ingredient commonly mixed with the pyrethroids is PBO (piperonyl butoxide). Alone, each substance has limited toxicity to insect species; when combined, the mixture is extremely toxic. PBO potentiates the pyrethroid by destroying one of the enzymes in the detoxification pathway that deactivates the pesticide in the insect. Humans exposed to this mixture suffer impaired ability of the liver to metabolize toxins in the environment.

Pesticides can be absorbed through the skin, inhaled, or swallowed. Many building products and household furnishings such as carpets, paints, and wood products are treated with biocides. The biocides enter the body through inhalation of fumes, skin absorption, or ingestion resultant from touching the contaminated product. Infants and small children are more likely to be harmed by pesticides because they are more sensitive and more likely to come into direct contact with treated carpets and lawns.

Pesticides can drift a long distance from the site of application, leaving residues throughout the surrounding community. Pesticides contaminate everything and everyone they contact. Residues are found in rain, fog, snow, food, water, livestock, wildlife, newborn babies, and even in the Arctic ice pack. People and pets may track pesticide residues into the house. An EPA study in Florida found the highest household pesticide residues in carpet dust.[8]

Pesticides may cause both acute and chronic health effects. Acute health effects appear shortly after exposure. Chronic health effects may not be apparent until months or years after exposure. Chronic effects generally result from long-term exposure to low levels of toxic chemicals, but may also arise from short-term exposure. A tragic misconception about pesticides is that the potential for harm is primarily the result of acute or immediate poisoning. In fact, delayed effects pose the greatest problems to human health. Many pesticides are fat soluble and bioaccumulate in tissues where they can exert prolonged effects on the immune, endocrine, and nervous systems. Children are most susceptible because their developing organs and nervous systems are more easily damaged.

Pesticide Facts

- A National Cancer Institute study indicated that the likelihood of a child contracting leukemia was more than six times greater in households where herbicides were used for lawn care.[9]
- According to a report in the *American Journal of Epidemiology*, more children with brain tumors and other cancers were found to have had exposure to insecticides than children without cancer.[10]
- According to the New York State Attorney General's office, 95% of the pesticides used on residential lawns are considered probable carcinogens by the EPA.[11]
- 2,4-D was a component of Agent Orange and is used in about 1,500 lawn care products.[12]
- Pesticides have been linked to the alarming rise in the rate of breast cancer.[13]
- Besides causing cancer, pesticides have the potential to cause infertility, birth defects, learning disorders, neurologi-

cal disorders, allergies, and multiple chemical sensitivities, among other disorders of the immune system.

When building or remodeling a healthy home, you can lower your pesticide exposure by not treating the soil under the building, and by eliminating standard building products that contain biocides. The "Division 10 - Specialties" section in this book includes discussion of pest management which emphasizes prevention of pest invasions through the use of construction techniques that create physical barriers.

Case Study:

Acute exposure to pesticides with long-term consequences

Louise Pape is a 55-year-old woman whose life changed drastically in 1993. On a warm spring day, she and her husband were slowly driving home with the windows rolled down to enjoy the cool breeze. At the roadside she spotted a man from a tree care company wearing a gas mask and spraying pesticides on the trees with a large hose. Louise suddenly felt a shower of chemicals on her face, in her eyes, nose, and mouth as the sprayer overshot his target. She later learned that the pesticide was a mixture of Malathion and Sevin.

The incident was the beginning of a nightmare illness for Louise, an environmental planner who, ironically, had just finished developing a safe pesticide plan for her employer, a transnational corporation. She was disabled for several months with flu-like symptoms, aching joints and muscles, severe headaches, dizziness, thyroid problems, insomnia, and shortness of breath. She was often bedridden and sometimes lapsed into a near comatose state upon re-exposure to even minute amounts of pesticides. Louise eventually developed full-blown multiple chemical sensitivity disorder. For the past four years, she has been virtually homebound, still unable to tolerate the trace amounts of pesticide and other chemical exposures that occur during routine activities when out in the world.

Her illness notwithstanding, Louise and her husband have become articulate spokespersons in educating the public regarding the hazards of pesticides and other chemicals. The ranch home they recently built, designed by Paula Baker, has become a model for nontoxic living.

Discussion

Many of the most harmful pesticides fall into three categories: organochlorines, organophosphates, and carbamates. In the above case, the onset of illness was associated with a single large exposure to an organophosphate and carbamate mixture. The cause of the prolonged illness was obvious. In most cases, however, the cause is not so obvious. Many people are exposed to repeated, low dose applications of pesticides which can result in general malaise with flu-like symptoms, chronic fatigue, and subtle neurological deficits. When patients complain of such symptoms to the doctor, they are rarely questioned about exposures to chemicals such as pesticides. Most emergency room doctors are familiar with acute pesticide poisoning, but few physicians have knowledge regarding long-term, chronic effects.

Case study: Chronic illness from "harmless" pesticide

Barbara Adler is a 43-year-old woman who was in good health until March 1996 when she developed the sudden onset of severe migraine headaches, loss of energy, frequent dizzy spells, and difficulty concentrating. Barbara consulted with a neurologist and many other health care practitioners over the ensuing months. None were able to help relieve her symptoms or shed light on the cause of her deteriorating health.

At some point in her search for wellness, Barbara reviewed the journal she had been keeping in which she recorded certain events in her life. She noted that around the time of the onset of her symptoms, her husband had purchased a bug spray from one of the local nurseries. He was told that the insecticide would be appropriate for the bugs on his houseplants. Barbara remembers that the bug spray smelled noxious to her, which prompted her to put some of the sprayed plants in the garage. She looked at the label on the bottle and saw that it contained Diazinon, a potent organophosphate known to have toxic effects on the nervous system. Barbara returned to the nursery to register a complaint and was told that Diazinon was not harmful.

Discussion

Although it is illegal for manufacturers to claim their pesticides are "safe," Dr. Elliott notes that in her experience local nurseries and other establishments selling pesticides frequently tell customers that organophosphates such as Sevin, Dursban, and Diazinon are harmless when applied according to instructions. In fact, many people with multiple chemical sensitivity disorder attribute the onset of their illness to pesticide exposure. While the patient in the above case became ill after an acute exposure to which she reacted immediately, the majority of cases occur after repeated, low-level exposures that can cause a gradual decline in health and vitality.

Case Study:
Chronic illness from repeated low-level exposure to pesticides

E. Merriam is a 48-year-old woman who complained of frequent flu-like symptoms since beginning employment at a new location. Symptoms seemed to recur every month and were especially severe over the winter. Conventional medications were of no benefit. After two years of watching her health decline, the patient discovered that the building in which she worked was being treated prophylactically one weekend a month with a pesticide that contained an organophosphate called Dursban. She then associated her flu-like symptoms with the monthly pesticide applications. The patient felt she could no longer continue to jeopardize her health and left her job. Three years later, she finally regained her health but continues to remain sensitive to petrochemicals.

Electromagnetic Field Pollution

Electromagnetic energy is ubiquitous. Some electromagnetic waves are natural, such as sunlight. Other fields are generated by human activity, such as radio and television waves, microwaves, and power line frequencies. Scientists classify EM waves according to frequency which corresponds to the wavelength. At one end of the spectrum are the infinitesimally short, high frequency gamma rays. At the other end are long, extremely low frequency vibrating waves used by submarines for underwater communication that may stretch for thousands of miles.

The magnetic field that envelops the Earth produces a steady, nonoscillating direct current at 7.83 cycles per second, or 7.83 Hertz, similar to that of the human body. This current pulsates on and off but always moves in a single direction. Each cell in the body has a pulsating vibration with an associated electromagnetic field. Communication between cells in the bodies is a function of electrical charges. These charges generate electrical currents which govern many of the body's major functions, such as heart beat, nerve conduction, and transport across cell membranes. These natural fields pulse on and off, but do not oscillate.

Manufactured fields oscillate back and forth. Unlike natural current, the fields change, and are thus called alternating current (AC). The electrical power grid operates at 60 Hertz (cycles per second) and simultaneously produces an electric and a magnetic field. Each field, electric and magnetic, has distinct properties and is measured separately using different meters. Common sources of 60 Hertz electromagnetic fields include power lines, electrical wires, electric blankets, fluorescent lights, televisions, and other household appliances.

Radio waves are a form of electromagnetic field. On a typical radio receiver set, you will find a dial used to change stations, known as the frequency tuner. As you move the dial to the higher numbers, you are increasing the frequency of the radio electromagnetic field you are seeking; when you move the dial to lower numbers, you are reducing the frequency. AM radio stations are found between 550 and 1600 kiloHertz (550,000 and 1,600,000 Hertz), or cycles per second. If the radio could be tuned all the way down the dial to 60 Hertz, you would be listening to the (very noisy) sound of electrical equipment.

There are major differences between radio waves and the electrical waves used to power electrical equipment. One difference is that radio waves are broadcast through the air; they are wireless and can travel for extended distances. It would be ideal if the fields from wiring systems stayed in the wires while the electricity is transported from one place to another. The problem is that wires "leak." The wires broadcast electromagnetic fields with the distance dependent on amperage. Amperage is analogous to the volume at which electric fields are being transmitted through the wires.

Assume that a radio receiver could be tuned to radio stations operating between 40 and 80 Hertz. As mentioned previously, if you were to tune the dial to 60 Hertz, you would hear a lot of noise because electricity in the United States operates at 60 Hertz. Next, assume that you take the same radio receiver to Europe. Upon tuning the radio to 60 Hertz, you would hear nothing because the European power system operates at 50 Hertz.

To experiment with the sound of electricity, take a cheap AM transistor radio and tune it between stations at the low end of the dial. By holding the radio near an operating electrical appliance or a dimmer switch you will pick up static or a buzzing noise. The noise you hear is not the 60

Hertz frequency, but rather higher frequency interference created by the 60 Hertz frequency.

Before meters for measuring elevated electromagnetic fields were readily available, some people would use an AM radio to obtain a rough approximation of whether an area contained elevated manufactured electromagnetic fields. The method is far from foolproof, but it was certainly better than nothing. Several models of inexpensive meters are available today which provide more accurate assessments.

Although long considered harmless, magnetic fields emitted by appliances and other electric objects are now suspected of promoting cancer, especially leukemia, lymphoma, and thyroid and brain cancer. Research has indicated that magnetic fields can induce a small electrical field inside the body, which in turn creates an electric current in and around the cells.[14] This current can alter the function of cell chemistry and can inhibit or enhance cell growth, ultimately affecting human health. Although there is no consistent dose/response relationship between magnetic fields and cancer, experiments on laboratory animals have shown that magnetic fields cause changes in protein synthesis and hormone levels. At present, the EPA refers to magnetic fields as possible carcinogens and continues its research.

The possibility of electromagnetic field influences on human health has been hotly debated for many years. Media attention has attempted to simplify the debate, reducing it a single question: Does exposure to EMF cause cancer? For over two decades the evidence has waffled back and forth. At present, reports of "no cancer effects" appear to dominate in the mainstream media. However, this is not an accurate representation. In recent years scientists have been refining research designs and approaches. Most scientists now recognize that it is difficult if not impossible to study groups of people that have experienced no exposure to any EM fields. Nearly all contemporary studies focus on differences between groups with some exposure versus groups with more exposure. Nearly everyone experiences EMF exposure.

Scientists are also recognizing that exposure to multiple fields may have the greatest influence on cancer incidence. The relationship of mixed electromagnetic fields to incidence of cancer is one of the most promising areas of research. A single field acting alone appears to have only minor or no effect, but when certain types of field exposure are simultaneously combined, the incidence of cancer increases dramatically.

Detailed explanations regarding the impact of EMF on human health are beyond the scope of this book. It is important to note, however, that for over 12 years over 300 top scientists from around the world gather in

November every year for a conference sponsored by the Department of Energy, the EPA, and the Electric Power Research Institute. For almost a week they present papers and review respective merits and faults. Nearly every scientist attending these meetings agrees that electromagnetic fields have biological effects, but they disagree on the exact effects and whether they are deleterious.

Studies in Europe have indicated that exposure to varying levels of electric fields can contribute to nervous disorders such as insomnia, depression, and anxiety. One U.S. study found an increase in aggressive behavior among baboons exposed to electric fields.[15] Recent studies have indicated that there may be a synergistic influence between magnetic and electric field exposures, making the combination more harmful than exposure to either field alone.

Over the past 50 years, people have been exposed to ever increasing amounts of manufactured radiation. The long-term consequences of this exposure are not clearly understood. Millions of Americans are now unwittingly engaged in long-term experiments on themselves.

Whereas Sweden has set limits for certain types of electric and magnetic field exposure, the U.S. government has not yet set similar standards because there is still no definitive information on hazardous levels. Given the enormous environmental stresses with which everyone is forced to cope, reducing exposure to these types of manufactured fields in homes becomes important in the attempt to reduce the overall stress load. Several simple and inexpensive measures can be incorporated into the design and construction of a new home that will reduce exposure, and at the same time reduce the risk of fire or electrocution. (See material under Divisions 2, 11, and 16 in this book.)

Interestingly enough, there are irrefutable reasons for controlling electromagnetic fields that have nothing to do with effects on human health. Fields can cause interference with entertainment equipment and computers. Electricians wiring hospital facilities must be extremely careful to use low field wiring techniques to prevent interference with critical care monitors and to prevent electric field sparking which could cause fires or explosions in oxygenated environments.

Curiously, the magnetic portion of the electromagnetic field is indirectly prohibited by the National Electric Code. Yet many new homes and even more older homes wired with standard techniques will have at least one area with an elevated field. This outcome is generally the result of a wiring error which violates code. The National Electric Code was established to help prevent fire, shock, and electrocution. Ask your electrician if it is acceptable for a structure's wiring to exhibit "net current." She or he will respond, "Absolutely not." Net current is what causes magnetic fields, and is one of the conditions that the structural wiring portion of this book seeks to avoid. (See "Division 16 - Electrical.")

Naturally Occurring Pollutants

Not all toxins are manufactured. Some naturally occurring substances in homes can have harmful effects on humans. Some of these pollutants include radon and radioactive contaminants, trace metals, house dust, molds, and pollens.

Radioactive Contaminants

Radioactive contaminants such as radium and uranium occur naturally within the Earth's crust. During the decay or breakdown of uranium, radon is produced. Radon is an invisible, odorless radioactive gas that seeps from the ground into homes, commonly through cracks in the foundation, basement slab, or through mechanical openings. Radon can also enter into the groundwater and affect water supplies.

Closed spaces present a hazard because radon levels can build up to values thousands of times higher than outdoor levels. High radon levels can cause radiation exposure equivalent to thousands of chest x-rays per person on an annual basis. Information on detecting and preventing radon contamination in homes is provided in "Division 7 - Thermal and Moisture Control."

Heavy Metals

Heavy metals in trace amounts can often be found in drinking water. These metals such as aluminum, copper, and lead can accumulate over time in human tissues, and are known to cause damage to the brain, liver, and kidneys. Having drinking water tested for contaminants is advisable; use a water purification device if necessary. Refer to "Division 11 - Equipment" for further information.

Biological Pollutants

Biological pollutants include pollen, house dust, and mold spores. Pollens from weeds, grasses, flowers, bushes, and trees enter the house through the doors and windows. They can be problematic for people with allergies. Air filtration methods are addressed in the "Division 15 - Mechanical."

House dust is composed of much more than simply soil. It is a complex mixture of dust mites, animal dander, mold spores, textile particles, heavy metals from car exhaust, skin cells, and more. Mites are a major culprit in causing allergies from house dust. They feed on skin cells and breed in mattresses, pillows, carpets, and upholstered furniture. Although generally harmless, their skeletal parts and fecal matter which stick to dust, can elicit allergic reactions in sensitive people.

Case Study:
Chronic illness due to acute exposure to virulent mold species

Tomasita Gallegos is a 37-year-old woman who first consulted Dr. Elliott in 1993. At that time she was frightened, in a state of severe agitation, and somewhat disoriented. Her face was bright red; her mouth showed increased salivation; her eyes were watery with constricted pupils; and her skin was warm to the touch. She was referred to Dr. Elliott by another physician who felt she might have experienced a pesticide exposure.

Ms. Gallegos was employed as a housekeeper in a private home. The morning of the day she became ill, Ms. Gallegos was instructed to open up a guest house which had been closed for a prolonged period of time. Shortly after the patient entered the guest house, she became acutely ill with the abovementioned symptoms. After the acute symptoms subsided, Ms. Gallegos was left with multiple problems, including chronic fatigue, panic attacks, chest pains, headaches, memory loss, and extreme chemical sensitivity. Her constellation of symptoms were baffling since it was determined that no harmful chemicals had been used on the premises.

An environmental engineering company was consulted to evaluate the guest house. Upon removing the furnace and cooling coils to allow access for a thorough cleaning of the ductwork system, the consultant found approximately two inches of water with green slime at the bottom of the supply plenum. Because the area was dark and cool and in the direct air stream of the house ductwork, the spread of microorganisms was very likely. Close inspection revealed that a defective humidification system was the source of the leaking water. Most of the microbial agents were fungi that, although found widely in nature, were highly concentrated in the interior environment. Many fungi produce toxic compounds called "mycotoxins." The intense microbial exposure had the effect of sensitizing the patient, leaving her with an overreactive immune system, known as "environmental illness."

At present, with diligent avoidance of molds, toxic chemicals, and allergens, Ms. Gallegos is slowly beginning to regain her health. Why was she so severely affected from such a brief exposure? The type of mold was a particularly virulent species. In addition, some individuals are more susceptible to toxins than others due to biochemical specificities. If the detoxification pathway in the liver is already at maximum capacity, then it might take only a relatively small exposure to overwhelm the system. This theory is called the "rain barrel effect" and refers to total toxic load. When more toxins enter the "barrel" than the body can excrete, then the barrel overflows and symptoms develop.

Straw bale residence featuring metal roofs, plaster finishes, tile and recycled wood flooring. (Architect: Paula Baker. Contractor: Prull and Associates, Inc. Photo: Julie Dean.)

Case Study: Asthma related to mold exposure

Dori Bennett is a 46-year-old woman who consulted with Dr. Elliott for the sudden onset of severe asthma. She had apparently been in good health until she moved into a new home. A leak in the home was repaired prior to the move, and the house had passed inspection. After her asthma progressed to the point of requiring hospitalization, it was suspected that the source of her problem was in her home. An environmental consulting firm noted heavy growth of mold in the crawl space. Molds found included aspergillus, actinomycetes, bacillus, cladosporium fusarium, mucor, penicillium, phoma, and ulocladium. Several strains of virulent molds grew on culture plates, some of which are known to cause asthma, pneumonia, hypersensitivity pneumonitis, and immune dysfunction.

Her family hired an environmental cleanup crew to rid the house of the mold in order to prepare it for resale. An outdoor unit was constructed to house a large heater fan to blow hot air under the house, while a unit on the opposite side of the house sucked out the air that had been blown in. It took six weeks to dry out the earth under the house. This procedure was followed by spraying hydrogen peroxide from a truck to saturate the earth. When the earth dried, the truck returned, this time with an iodine solution that was sprayed under the house to kill the mold and retard further growth. The crew then donned what looked like spacesuits and washed the entire house interior with hydrogen peroxide and water in a 1:1 mixture. Every item in the house was removed and vacuumed with an HEPA filter and then cleaned. Items that could not be cleaned were thrown away. The cost of the cleanup was $40,000. When the house was certified to be free of mold, it was sold. Ms. Bennett now lives in a mold-free home and her health is slowly improving.

Mold plays a significant role in triggering allergies, asthma, and chemical sensitivities. Mold can produce by-products as toxic as some of the most hazardous manufactured chemicals that affect the nervous and immune systems. Mold is commonly as-sumed to be found only in older homes, but it can be found wherever moisture accumulates, such as in basements, bathrooms, window sills, laundry rooms, or wherever leaks and flooding occur. Moist building materials, including new materials, can become breeding grounds for mold and bacteria within a few days. Many of the materials used in standard construction of new homes are susceptible to water damage and fungal growth. A moldy home is frequently a sign of a home with deteriorating building materials. Even when molds are contained inside walls or other building cavities such as attics and crawl spaces, the slightest air current can send fungal spores swirling through the air where they can be easily inhaled.

Carpets act as large reservoirs for dust, bacteria, and mold. Microbes commonly grow within the ductwork of forced air heating systems which can result in mold and dust spread throughout the house. Unless kept spotlessly clean, toilets and many modern appliances which use water reservoirs such as vaporizers and humidifiers, can breed microbes. Methods for preventing and controlling mold infestation are discussed throughout the specifications.

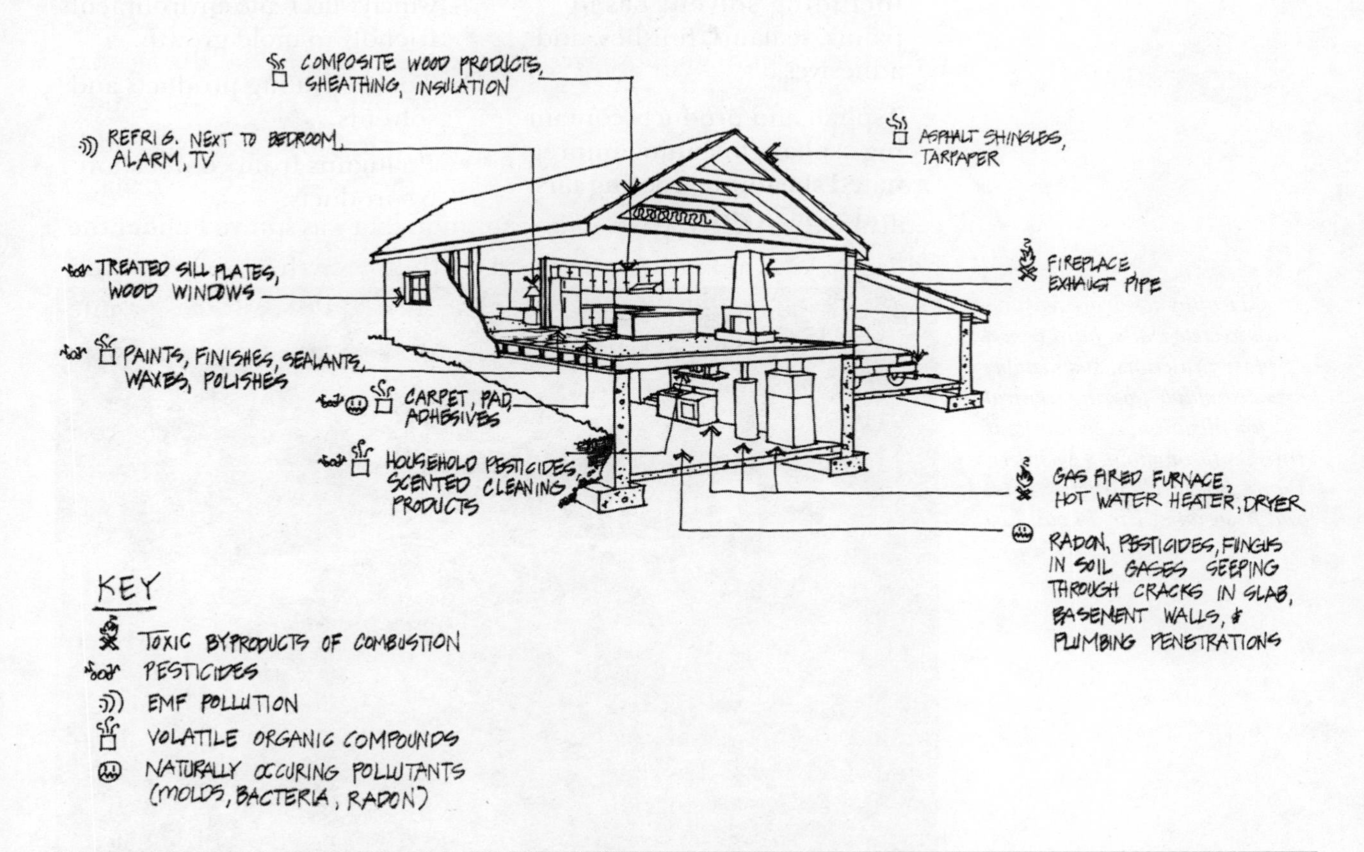

Illustration of indoor pollution sources.

Common Sources of Pollution in Construction

The use of toxic substances in construction is standard, and unless you specifically ask to have these substances eliminated from your project, they will be included. The book is organized in the order of standard building specifications which are frequently used by architects and builders. We will explain the many instances, some obvious and others less so, where undesirable materials and practices might be found. Healthier options from which you can choose are then listed. As a brief overview, the products and practices listed below are the most common sources of pollution in new construction.

- Insecticides, mildewcides, herbicides and other biocides found in building materials and applied on site.
- Composite wood products (chemically treated woods), including particleboard, chipboard, plywood, sill plates, and manufactured sheathing.

- Building products and additives that emit harmful VOCs, including solvent based paints, sealants, finishes, and adhesives.
- Asphalt and products containing asphalt, including impregnated sheathing, roofing tars, and asphalt driveways.
- Building materials containing mold, mildew, or practices which will create environments friendly to mold growth.
- Toxic cleaning products and solvents.
- Pollutants from combustion by-products.

Healthy residence features pumicecrete walls, hard trowel plaster interiors, sustainably harvested maple flooring, central air filtration, and nontoxic finishes throughout. (Architect: Paula Baker. Contractor: Prull and Associates, Inc. Photo: Lisl Dennis.)

Case Study: The relationship between allergies and chemical exposure

In the 1950s it was estimated that about 14% of the population suffered from allergies. According to some estimates, this proportion at present is estimated at between 40 and 75%. Why the dramatic increase? Allergists in Japan pondered the same question. A hypothesis was put forth that certain chemicals act as sensitizing agents. To test the hypothesis, two groups of mice were exposed to high levels of the Japanese equivalent of juniper pollen, and then tested for an allergic response. In both the study and the control group about 5% of the mice developed allergies to the pollen. The study group was then exposed to benzene fumes from car exhaust. Upon retesting, there was a significant increase in the study group's allergic response to the pollen while the control group remained at 5%. The experiment is described by M. Muranaka, et al., in "Adjutant activity of diesel exhaust particulates for the production of IgE antibody in mice," in the *Journal of Allergy and Clinical Immunology* (Vol. 77, April 1986, 616-623).

Discussion

Although there is clearly a link between chemical exposures and allergies, the exact mechanism has not yet been elucidated. Most people who have acquired multiple chemical sensitivities also suffer from traditional allergies to pollens, dust, dander, and mold.

Benzene is only one of many pollutants known to damage the immune system. These chemicals are found in thousands of modern products for home and industrial use, and therefore, millions of people are constantly exposed to low levels of these chemicals at work and at home.

Strategies for Source Reduction of Pollutants

As explained previously, modern building techniques have created sick buildings by increasing the sources of indoor pollution while decreasing the rate of air exchange. It stands to reason that in order to make buildings healthy environments, we must find ways to reduce the sources of pollution while introducing more air exchange. Strategies for achieving source reduction are organized in order of effectiveness below.

1. Eliminate sources of pollution.
2. Substitute healthier materials.
3. Exercise prudence when using unavoidable toxic substances.
4. Cure materials before they are installed within the building envelope.
5. Seal materials so that they offgas less.

Elimination

If all toxins could simply be eliminated from buildings, we would have the basis for an ideal environment. In many situations, this is not only possible, but also cost effective. For example, countertop materials can often be attached to cabinets with mechanical fasteners, thereby eliminating the need for toxic adhesives.

Substitution

Where chemicals must be used, then it is almost always possible to substitute a less toxic substance in place of a standard one. For example, paint with no VOCs or preservatives can be specified in substitution for a standard paint which contains harmful chemicals such as formaldehyde.

Prudent Use

In a few cases, the use of a toxic substance is unavoidable. For example, there is no acceptable substitute for the solvent based glues used to join plastic plumbing lines. However, the specifications outlined in the book provide guidelines to reduce the amount of adhesive used.

Curing

In some cases where toxic substances are chosen for reasons such as cost or durability, the impact of the product will be reduced if it is properly cured. In the specifications, for example, curing plywood before application is explained. Many materials can be purchased with factory applied finishes that have been heat cured. Such finishes, which may have been quite noxious in their liquid state, are safely applied and cured under controlled conditions. Many factory applied finishes will have little impact on air quality by the time they are installed in the home.

Sealing

If a toxic building component cannot be eliminated or substituted, then sealing it will help to reduce the rate of offgassing. Although this approach is far from perfect, there are cases where we recommend vapor sealants or barriers for this purpose. For example, premanufactured wood windows usually contain pesticides. As it is almost always cost-prohibitive to have custom windows made, sealing the windows with a clear sealer will help limit pesticide exposure.

Throughout the planning of a healthy home, you will be weighing the health risks, cost, time, and aesthetics of the above five strategies to find the solution that is best for you.

Fragrances in the Home

Artificial fragrances are found throughout most homes and workplaces in body and hair products, household cleaners, detergents, fabric softeners, air fresheners, and even in some magazines. Fragrance is cited as an indoor irritant and pollutant in several major studies, including the EPA's TEAM Study (Total Exposure Assessment Methodology Study, June 1987), and "Chemical Sensitivity: A Report to the New Jersey State Department of Health," by Nicholas Ashfor, Ph.D. and Claudia S. Miller, M.D. (December 1989).

In the days before "better living through chemistry," fragrances were made from flowers. Now, approximately 95% of all ingredients used by the fragrance industry are synthetic.[16] According to the U.S. Food and Drug Administration, about 4,000 petroleum derived chemicals are used in fragrances.[17] These include toluene, formaldehyde, acetone, benzene derivatives, methylene chloride, phenyl ethyl alcohol, methyl ethyl ketone, and benzyl acetate. A single fragrance can contain as many as 600 different chemicals.

In a 1988 study, the National Institute of Occupational Safety and Health found that in a partial list of 2,983 chemicals now being used by the fragrance industry, 884 toxic substances were identified.[18] Many of these substances are capable of causing cancer, birth defects, central nervous system disorders, reproductive disorders, and skin irritation. According to the National Academy of Science, there is minimal or no data on toxicity on 84% of the ingredients found in fragrances.[19]

Currently there is no agency regulating the fragrance industry. The FDA is aware of the serious nature of the problem but is unable to undertake the astronomical expense of testing each of the hundreds of chemicals found in fragrances. Without such testing, the FDA would be subject to lawsuits by manufacturers if fragrances were banned. Thus, as is often the case, the onus falls on the consumer to make informed choices.

If you create a healthy house, it seems senseless to introduce artificial fragrances which will sabotage your efforts to breathe clean air. In the resource list, we have compiled names of companies that supply fragrance free products, or products with fragrances derived from natural sources.

Case Study:
Fragrant fumes

E.B. is a 58-year-old man with a ten-year history of chronic sinus congestion, hoarseness, and headaches. By the time he consulted with Dr. Elliott, he had tried many forms of treatment including nasal surgery, frequent courses of antibiotics, decongestants, and steroid nasal drops. After removing dairy products from his diet, he noticed only a partial improvement in the congestion. Dr. Elliott then suggested that he try eliminating all scented products from his body, including detergents, soaps, and colognes. Through a process of trial and error, E.B. discovered that his aftershave lotion was a significant cause of his symptoms. His voice has now returned to its former resonance and he is without headaches or sinus congestion.

Discussion

Millions of people are made ill by artificial fragrances. Most people are unaware that fragrances can cause or contribute to health problems. The most common symptoms related to fragrances include asthma, headaches, dizziness, fatigue, mental confusion, memory loss, nausea, irritability, depression, rashes, and muscle and joint pains. With increasing awareness and growing demand, products are now available to the public that are fragrance free or scented with purely plant derived substances.

Regional Differences in Climate

The building industry in the United States, a country with vast climatic variations, is primarily regulated by a handful of building codes. These codes do not sufficiently address the fact that each climatic zone carries particular concerns as to how moisture, temperature, wind, vegetation, and wildlife will impact the building envelope. Historically, regional building types throughout the world evolved over time whereby local materials were fashioned into a perfect response for the surrounding climatic conditions. With the advent of mass produced, standardized building materials and housing, much of this indigenous wisdom has been lost. A new suburban home in Cincinnati, for example, may look identical to one built in Los Angeles. In spite of the vastly different climatic conditions in Cincinnati and Los Angeles, the two buildings will be mechanically equipped to provide the occupant with interior conditions of 70 degrees Fahrenheit 24 hours a day, 365 days a year.

However, the interaction between the climate and the building envelope in the previous two examples will be very different. Residential building techniques have undergone sweeping experimentation since WWII with the introduction of mass produced and transported building components. Professionals in the building industry are now discovering that certain assumptions made 20 years ago about how the new products interact with climate were shortsighted. Many of the new products result in buildings that do not support human health or use energy efficiently.

It is essential that architects, builders, and homeowners become familiar with the localized conditions of the potential home site. An inquiry into the kinds of problems that have developed in local buildings due to the natural environment would be a beneficial undertaking. The local building lore can potentially be a rich source of information. Listed below are a few examples (by no means exclusive) of differing regional conditions and respective challenges.

- The air of coastal locations is typically characterized by high salt content, resulting in metal corrosion.
- Areas experiencing alternating freeze/thaw conditions will be subject to ice damming problems. Buildings will also be much more susceptible to deterioration caused by water seeping into cracks and then expanding as it turns to ice.
- Wood products exposed to the elements in southwestern deserts will suffer from accelerated UV exposure and drying.
- Moisture and mold problems associated with condensation caused by air conditioning are typical in climates with high temperatures and humidity.
- Fire safety is a major concern in wooded areas.
- Nearly every region is specific in terms of insect and vermin problems.

In summary, certain problems unique to your building location will not be remedied or addressed by building codes or materials manufacturers. Architects, builders, and homeowners must be jointly responsible for investigating specificities.

Endnotes

1. National Research Council, Assembly of Life Sciences, *Indoor Pollutants* (National Academy Press, 1986). Cited in *The Human Consequences of the Chemical Problem* by Cindy Duehring and Cynthia Wilson (White Sulphur Springs, MT: Chemical Injury Information Network, 1994), 3.
2. "Environment 1992," *Science News* 142:25-26 (December 17 and 26, 1992), 436.
3. William Rea, *Chemical Sensitivity*, Vol. 2 (Lewis Publishers, 1994), p. 706.
4. Marion Moses, *Designer Poisons* (San Francisco, CA: Pesticide Education Center, 1995), p. 309.
5. U.S. General Accounting Office, "Lawn Care Pesticides: Risks Remain Uncertain While Prohibited Safety Claims Continue" (U.S. Government Printing Office, March 1990), 4-5.
6. Michael H. Surgan, "EPA Pesticide Registration: Our Safety in the Balance?" *NYCAP News* 4:3 (Fall 1993), 21-23.
7. Steven Arnold, et al., "Synergistic Activation of Estrogen Receptor with Combinations of Environmental Chemicals," *Science* 272 (June 7, 1996), 1489-1492.
8. Marcia Nishioka, et al., "Measuring Transport of Lawn-applied Herbicides from Turf to Home: Correlation of Dislodgeable 2,4-D Turf Residues with Carpet Dust and Carpet Surfaces Residues," *Environmental Science Technology* 30:1, 3313-3320.

9. Jack Leiss and David Savitz. "Home Pesticide Use and Childhood Cancer: A Case-control Study." *American Journal of Public Health* (February 1995), 249-252.
10. E. Gold, et al., "Risk Factors for Brain Tumors in Children," *American Journal of Epidemiology* 109 (1979), 309-319.
11. American Cancer Society. "Drug-free Lawns" (Pamphlet), 1993.
12. Rea, *Chemical Sensitivity*, p. 880.
13. D.J. Hunter and K.T. Kelsey, "Pesticide Residues and Breast Cancer: The Harvest of a Silent Spring?" *Journal of the National Cancer Institute* 85 (April 21, 1993), 598-599.
14. Robert Becker, *Body Electric* (Tarcher Press, Inc., 1985).
15. Research conducted by Anthony Coehlo and Steve Easley of Southwest Research Institute, San Antonio, TX. Cited in *Sacramento Bee* (July 14, 1990).
16. Report by the Committee on Science and Technology, U.S. House of Representatives (Report 98-821, September 16, 1986). Cited in Duehring and Wilson, *The Human Consequences of the Chemical Problem*, p. 5.
17. "Neurotoxins: At Home and the Workplace," Report by the Committee on Science and Technology, U.S. House of Representatives (Report 99-827, September 16, 1986). Cited in *Ibid.*
18. Neurotoxins: At Home and the Workplace," U.S. House of Representatives, Committee on Science and Technology, 99th Congress, 2nd Session (Report 99-827, June 1986). Cited in *Ibid.*
19. *Ibid.*

Further Reading and Resources

Publications

Anderson, Nina, et al. *Your Health and Your House.* Keats Publishing, 1995. A resource guide to health symptoms and the indoor air pollutants that aggravate them.

Bower, John. *Healthy House Building, A Design and Construction Guide.* The Healthy House Institute, 1993. Step-by-step guide illustrating author's construction of a model healthy house.

Bower, John. *The Healthy House: How to Buy One, How to Build One, How to Cure a 'Sick' One.* Lyle Stuart, 1989. Describes in great depth a three-step approach consisting of elimination, isolation, and ventilation. As many toxins as possible are identified and eliminated; a tight air barrier isolates occupants from infiltration; and air is exchanged and purified by means of mechanical ventilation. The author speaks from firsthand experience in successfully creating a chemical free sanctuary for his spouse.

Breecher, Maury M. and Shirley Linde. *Healthy Homes in a Toxic World.* John Wiley and Sons, 1992. The authors identify household health hazards, the human health conditions associated with them, and solutions for healthier environments.

Colburn, Theo, Dianne Dumanoski, and John Peterson Myers. *Our Stolen Future.* Plume, 1997. A gripping account of the scientific research linking reproductive failures, birth defects, and sexual abnormalities to synthetic chemicals that mimic natural hormones, causing disruption of the endocrime system.

Dadd, Debra. *Non-toxic, Natural, and Earthwise.* J.P. Tarcher, 1990. A practical, easy-to-use guide to nontoxic alternatives for cleaning products, personal care products, lawn and garden supplies, baby care items, pet care, and household furnishings.

The Green Guide. Address: 40 West 20th St., New York, NY 10011-4211. Tel: (888)ECO-INFO. This newsletter discusses various relevant topics and promotes safe and ecologically sound consumer choices.

Green, Nancy Sokol. *Poisoning Our Children.* The Noble Press, 1991. The contemporary pesticide problem comes alive as the author relates the nightmare she endured after unwittingly poisoning herself in her own home with repeated pesticide exposures.

Institute of Baubiologie and Ecology Correspondence Course. Available through Helmut Ziere, IBE, Box 387, Clearwater FL 34615, Tel: (813)461-4371. This certified home study course has been translated into English from the original work of Dr. Anton Schneider, the founder of baubiology. The course provides a comprehensive discussion of the interrelationship between the built environment, human health, and planetary ecology.

Lawson, Lynn. *Staying Well in a Toxic World: Understanding Environmental Illness, Multiple Chemical Sensitivities, Chemical Injuries, and Sick Building Syndrome.* Lynnword Press, 1994. Highly readable, informative, and comprehensive overview of the devastating effect of toxic surroundings, authored by a former medical writer with a thorough understanding of the contemporary chemical problem.

LeClaire, Kim and David Rousseau. *Environmental by Design: Interiors, A Sourcebook of Environmentally Aware Choices.* Hartley and Marks, 1993. Provides a "cradle to grave" environmental analysis of common building materials.

Our Toxic Times. Published by the Chemical Injury Information Network. Address: P.O. Box 301, White Sulphur Springs, MT 59645. Tel: (406)547-2255. A useful newsletter for people interested in understanding how chemicals impact human health.

Pearson, David. *The Natural House Book: Creating a Healthy, Harmonious, and Ecologically Sound Home Environment.* Fireside, 1989. The author gives a thoughtful explanation of the problems associated with standard building practices in terms of human health and environmental impacts. He then shows an inspiring array of natural building materials and systems from around the world.

Rogers, Sherry A., M.D. *Tired or Toxic.* Syracuse, NY: Prestige Publishing, 1990. Detailed and comprehensive medical explanations about how chemicals are impacting human health.

Roodman, David Malin and Nicholas Lenssen. *A Building Revolution: How Ecology and Health Concerns Are Transforming Construction.* Worldwatch Paper 124, March 1991.

Schoemaker, Joyce and Cherity Vitale. *Healthy Homes, Healthy Kids.* Island Press, 1991. The authors discuss ways to protect children from everyday environmental hazards found in the home.

Thrasher, Jack and Alan Broughton, *The Poisoning of Our Homes and Workplaces: The Indoor Formaldehyde Crisis.* Seadora, Inc., 1989. Detailed analysis of the indoor formaldehyde crisis in the United States.

Venolia, Carol. *Healing Environments, Your Guide to Indoor Well-being.* Celestial Arts, 1988. The author takes the reader through a series of environmental awareness raising exercises expanding a holistic approach to health and the built environment that includes the wellness of body, mind, and spirit.

Wilson, Cynthia. *Chemical Exposure and Human Health.* Jefferson, North Carolina: McFarland & Company, 1993. A reference guide to 314 chemicals with a list of symptoms they can produce, and a directory of organizations.

Zamm, Alfred and Robert Gannon. *Why Your Home May Endanger Your Health.* Simon and Schuster, 1982. Based on a ten-year scientific study, this book explains how millions of Americans may be suffering ill health because their homes have become toxic chambers. The author discusses remedies for many of the major health hazards found in the home.

Nontoxic products

The Allergy Relief Shop, 3371 Whittle Springs Road, Knoxville, TN 37917. Tel: (800)626-2810. Mail order catalog offering supplies and building products for the allergy free home.

Allergy Resources, P. O. Box 888, 264 Brookridge, Palmer Lake, CO 80133. Tel: (800)873-3529; (719)488-3630. Nontoxic cleaning compounds and body care products.

American Environmental Health Foundation, 8345 Walnut Hill Lane, Suite 225, Dallas, TX 75231. Tel: (800)428-2343, (214)361-9515. Sells a wide range of household, building, personal care, and medical products as well as organic clothing, books, and vitamins.

Aubreys Organics, 4419 N. Manhattan Ave., Tampa, FL 33614. Tel: (800)282-7394. Over 150 hair, skin, and body care products made from herbs and vitamins, without synthetic chemicals.

Building for Health Materials
Center, P.O. Box 113, Carbondale, CO 81623, Tel: (970)963-0437. For orders only: (800)292-4838. Distributor of a wide variety of healthy building products. The owner, Cedar Rose, is also a building contractor who has practical experience with most products sold by the Center.

Dasun Company, P.O. Box 668, Escondido, CA 92033. Tel: (800) 433-8929. Catalog sales of air and water purification products.

Eco Products Inc., 1780 55th Street, Boulder, CO 80301. Tel: (303) 449-1876. Supplier of ecologically sound building products.

The Living Source, P.O. Box 20155, Waco, TX 76702. Tel: (817) 776-4878. Voice mail order line: (800) 662-8787. Catalog sales of "products for the environmentally aware and chemically sensitive."

The Natural Choice, 1365 Rufina Circle, Santa Fe, NM 87505. Tel: (800) 621-2591. Catalog sales of natural paints, stains, and healthy home products.

NEEDS, 527 Charles Ave., Suite 12A, Syracuse, NY 13209. Tel: (800) 634-1380. Mail order service offering a wide array of personal care products for the chemically sensitive.

Nontoxic Environments Catalog, P.O. Box 384, New Market, NH 03857. Tel: (603) 659-5919. For orders only: (800) 789-4348. Mail order catalog offering a wide range of supplies and building products for a nontoxic home.

The Nontoxic Hot Line. For consultations, phone (510) 472-8868. Line for orders (only): (800) 968-9355. Catalog sales of products for achieving and maintaining indoor air quality and safety for homes, offices, and automobiles.

Planetary Solutions, 2030 17th St., P.O. Box 1049, Boulder, CO 80302. Tel: (303) 442-6228. Environmentally sound materials for interiors.

Part II Specification

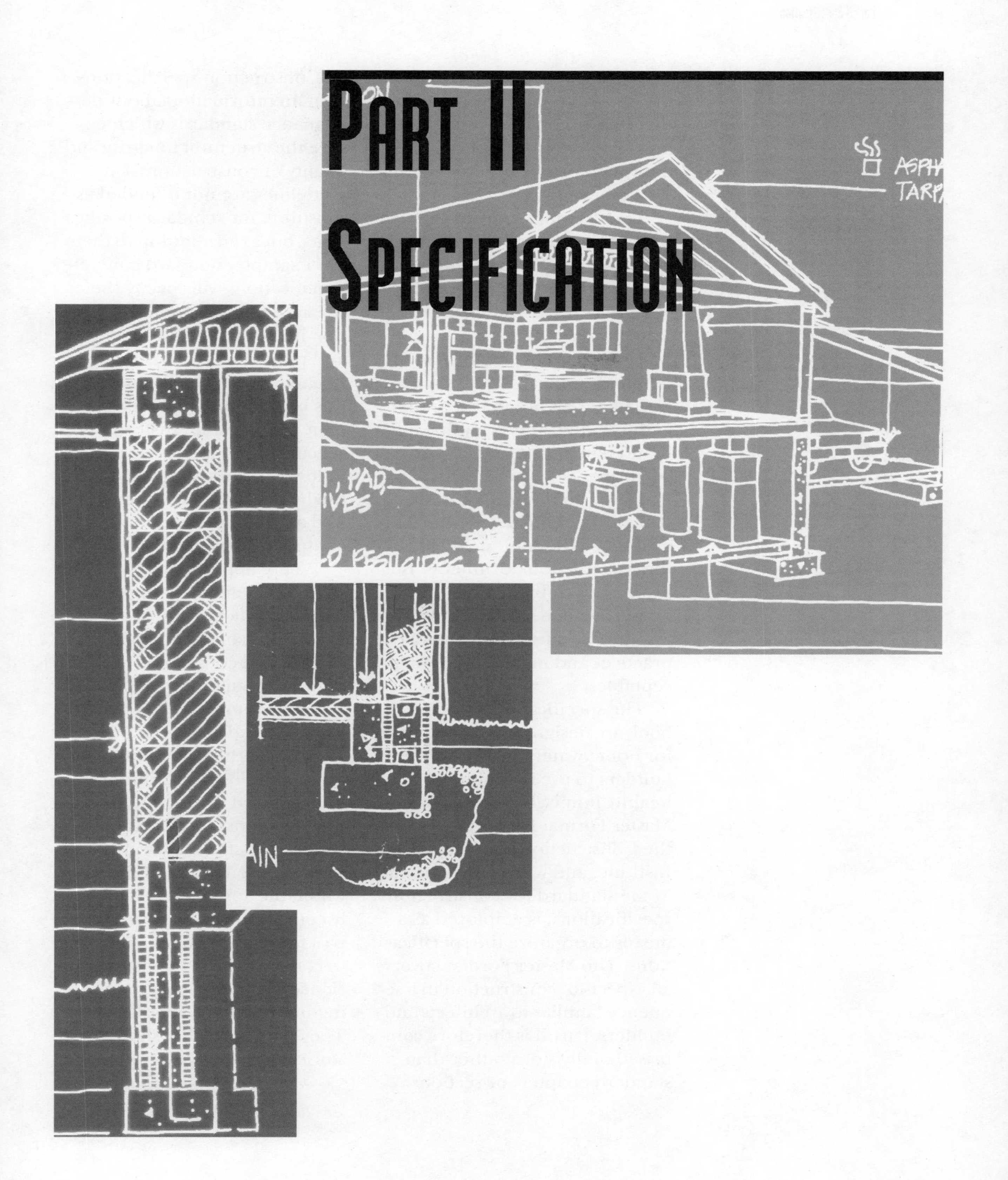

Construction specifications are the detailed written instructions that support architectural drawings. Together the "specs" and drawings comprise construction documents.

The drawings explain the physical layout and appearance of the building, how it will be structured, and the choice of general construction materials. Specifications contain instructions that cannot be shown easily on the drawings. They indicate how materials are to be handled and installed, and prescribe brand names of products and performance requirements.

For custom residential construction, a set of detailed specifications is often not included in the construction documents. But if you want to build a healthy home, detailed specifications are essential because many standard practices and materials are unacceptable.

The specifications in this book are designed as a guideline for homeowners, architects, and builders to use in building a healthy home. The 16 division Master Format list developed by the Construction Specification Institute, and widely recognized as the standard for construction specifications, is employed as a means to organize the specifications. The Master Format covers all aspects of construction in a sequence familiar to architects and builders. Part II is therefore comprised of divisions rather than standard chapters or sections.

Construction specifications contain information about performance standards which ensure the structural integrity and quality of construction. The guidelines are not intended as a substitute for standard specifications, but as an addition to them. For example, standard concrete specifications will specify the strength of concrete to be used, how it is to be mixed and poured, and procedures for testing its strength. The specifications in this book do not include such basic information. Instead, the specifications appearing in the following 16 divisions focus on the health of home occupants as well as of home builders and component installers.

Where appropriate, the differences between healthy and standard construction are explained. Products, manufacturers, tradespeople, and consultants involved in healthy building are specified. Toll-free numbers in the resource list are included when available so that you may conveniently locate the closest distributors.

Dispersed throughout the specifications are medical case studies, the stories of real people from different walks of life with whom the authors have personally come into contact over the past few years. What they all have in common is the firsthand experience of the consequences of living in unhealthy environments. They have agreed to share their stories with you.

Division 1 - General Requirements

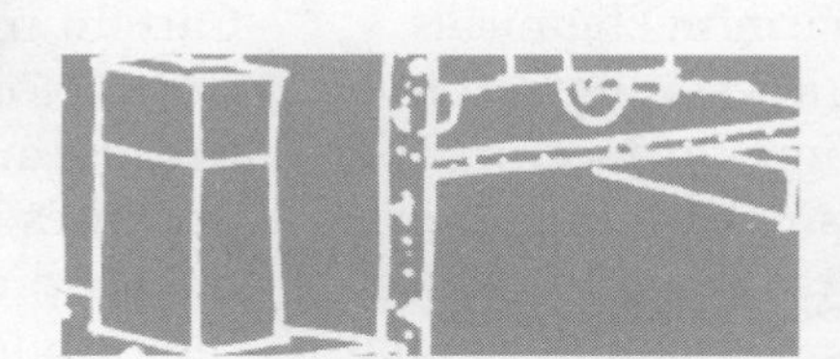

Clear communication among contractor, owner, and architect is a key factor in the success of any building project. When creating a healthy home, there are many special project procedures that must be communicated with even greater clarity than in standard construction. The owner's intentions and instructions for special projects can be formally transmitted in this division of the specifications, thus making them part of the construction contract.

Appearing below is sample specifications language that succinctly states what the owner wishes to create in a special project.

Statement of Intent

This house is being constructed as a healthy house. The products specified herein are intended to be as free of harmful chemicals as are presently available and reasonably attainable. In using these products, we are safeguarding, to the best of our ability, the health of future inhabitants, workers involved in this construction, workers involved in the manufacture of these products, and the planet.

Coordination

Building a healthy home can be a pioneering endeavor. Choosing the right architect and contractor for this task is of paramount importance. Creativity, intelligence, common sense, and enrollment in the ideals of healthy house building are essential characteristics. At times you and your team will be experimenting with products that have not been on the market long enough to have a performance history or wide distribution. At other times you may find yourself participating in a revival of materials and techniques that were used successfully for centuries but have been replaced in standard construction by commercialized products containing harmful chemicals.

The contractor will need to allow more time for locating special materials, scheduling their use, and supervising their installation. You may encounter initial resistance from subcontractors who are reluctant to do things that are unfamiliar. Some of the healthier products might be harder to work with because they do not contain certain additives which make application easier. For these reasons the general contractor will need to supervise the project more closely than in standard construction.

During the construction of a healthy building, the owner may wish to hire subcontractors to carry out specific environmentally related testing or installation. Included in standard contracts from the American Institute for Architecture (A.I.A.) is document A201, "General Conditions for Construction" which acknowledges the owner's right to hire her own subcontractors. Careful coordination with the contractor is necessary, however, because any delays and associated expenses incurred by the contractor due to this work will be the responsibility of the owner. Some of the additional testing and inspections are described in "Division 13 - Special Construction."

In summary, a healthy home can take more time and effort to build, which may be reflected in the contractor's scheduling and pricing. Once committed to the project, however, the contractor's role is like that of the symphony conductor who must lead all players to a successful performance regardless of the difficulty of the piece. You may wish to clearly state this expectation in your specifications.

Coordination

The contractor shall be responsible for obtaining all specified materials or approved substitutes and for performing all special project procedures within the contract time, as stated within the construction contract.

The contractor shall be responsible for the general performance of the subcontractors and tradespeople, and any necessary training, specifically with regards to the special project procedures, materials, and prohibitions as outlined in these specifications.

Special Project Procedures

Healthy home building does not permit many behaviors and practices that are commonly accepted at standard job sites. The procedural expectations must be clearly stated by the owner and upheld by the contractor. Appearing below are some basic rules that you may wish to include and expand upon in your specifications.

Special Project Procedures

The following special project procedures must be obeyed at all times:

- Smoking is prohibited within or near any structure on the job site.
- The use of gas-generated machinery and gas or kerosene fired heaters is prohibited within or near the building.
- No insecticides, herbicides, or chemicals other than those specified may be used on the job site without prior approval by architect/owner.
- All materials are to be protected from moisture damage during storage.

Special Site Procedures to Prevent Insect and Rodent Infestation

Some simple measures can be applied from the outset of construction that will prevent infestation of rodents and insects. Consider adding the following requirements to the above list.

Special Site Procedures

All foodstuffs shall be disposed of in containers which will be removed from the job site and emptied at the end of each workday.

All debris shall be removed from under and around the building premises and properly disposed of in a dumpster. The dumpster shall be removed when full on a regular basis so that piles of debris do not accumulate on the ground around it.

Quality Control

There may be some instances where you will be asked by your contractor to share in the responsibility for application of an experimental material. You may choose to accept this responsibility on a case by case basis. However, aside from any agreed upon exceptions, the contractor must be willing to provide the same warranty for your finished home that she or he would in using standard products. He should have no problem doing so as long as he carefully follows the manufacturer's instructions.

Manufacturers will specify the conditions required for the proper application of their products, such as proper curing times, acceptable temperature ranges, or specific preparation of substrates. Because the materials contain fewer chemical additives, the manufacturer's specifications may be both different and less forgiving than what tradespeople are accustomed to. Consequently, they may require close supervision by the contractor to maintain a high quality standard.

Another area requiring special vigilance on the contractor's part lies in the careful screening of materials as they arrive on site to ensure that no spoilage, absorption of odors, mold, or other forms of contamination have occurred.

We are aware of a case where batt insulation was shipped to the job site in a truck that was also used to transport fertilizer. Once installed, the home took on a distinctly unpleasant odor due to the absorbent nature of the batt insulation. In another case the painter, who was unfamiliar with milk based paints, did not realize that the products he was using had spoiled. As a result, the finished home smelled like rancid milk.

Subcontractors may be unfamiliar with some of the healthier products we recommend and thus may not recognize a problem when it arises. These products typically have little odor. A product that emits a strong odor may be an indication for concern. The contractor's nose becomes an important quality control mechanism.

Exceptions include products such as silicone caulks and vapor barrier sealants that have a strong odor upon application, but which quickly dissipate and become neutral. A call to the architect or manufacturer may be helpful for reassurance when questions arise.

It is important to have a clear agreement from the outset about your expectations concerning quality. This agreement can be formalized in the specifications. Sample language that may be added to your building specifications appears below.

Quality Control

The contractor shall perform and maintain the special project procedures with the same quality of workmanship as would be expected with standard materials and methods.

The contractor shall maintain a quality control program which ensures full protection of work against exposure to
prohibited materials and practices.

Except as otherwise approved by the architect or owner, the contractor shall determine and comply with the manufacturer's recommendations on product handling, storage, and protection.

The contractor shall verify that, prior to installation, all materials are undamaged, uncontaminated, and free of acquired odors. Any products found to be defective shall not be used unless approved by the owner/architect.

Signage

Even if your contractor is well aware of your intentions, he probably will not have the chance to personally speak to the hundreds of people who will work at your site during the course of construction. It is important that the special rules that apply to your home be posted in a prominent spot where all who enter will read them. You can specify that a job sign be made and posted. Sample wording that can be placed in your specifications with regards to signage follows.

Signage

The following sign is to be made and prominently posted on the job site. It is the responsibility of the general contractor to ensure that his labor force, all subcontractors and their labor forces, and all suppliers be made aware of these rules and follow them at all times.

Sign to be Posted

This house is being constructed as a healthy home. Only specified products and procedures may be used. If in doubt, contact the general contractor.

The use of any toxic substances such as pesticides, fungicides, or noxious cleaning products is prohibited anywhere on this site.

Smoking within or near the building and its garage or outbuildings is strictly prohibited.

No gasoline generated machines or open combustion heaters shall be used inside or near the house after the foundation is completed.

Spills of fuels, solvents, or chemicals must be avoided. If a spill occurs, report it to the general contractor immediately.

Alternatives to specified materials must be approved in writing by the owner and/or architect prior to use.

Case Study:
A mishandled spill

Early in his career as an environmental consultant, John Banta received a frantic call from a woman with chemical sensitivities who was in the process of having a home built. The client had painstakingly detailed plans and specifications with the help of John and her architect. The project had proceeded virtually without problems, and was entering the final interior painting and sealing process when a worker for the subcontracting painter accidentally kicked over a bucket of nontoxic paint, spilling it on the unfinished floor. The worker ran to his truck and grabbed a can of mineral spirits which he used to clean up the spill. The solvent soaked into the floor and the fumes filled the house. John's client became distraught because her new house was making her feel sick.

Discussion

In spite of all the best efforts, accidents still happen. The subcontractor failed to educate his worker. The spilled paint was water based which meant that the use of mineral spirits was unnecessary and inappropriate. The painter should have wiped up as much as possible using clean rags, and then scrubbed the rest with water. Since the floor was unfinished, any remaining paint could have been removed with sanding.

Many things were attempted to remove the noxious mineral spirit odor from the home, but the solvent had been absorbed by the construction materials. Even pulling up the contaminated portion of the floor was insufficient to fix the problem. The cleaning substance used by the painter was clearly in violation of the job contract and it appeared that a lawsuit was imminent. Fortunately, the house was quickly sold to a less sensitive person who wanted an ecologically constructed home and who was not affected by the residual odor of mineral spirits.

Prohibited Products

Because it is difficult to foresee every single product application that will be required in a project, listing the major categories of prohibited materials that are the worst health offenders is prudent. Sample language containing such a list follows.

Prohibited Products

The use of the following substances is prohibited:

- Herbicides, fungicides, and pesticides such as insecticides, except as specified
- Composite wood products containing ureaformaldahyde binders
- Asphalt or products containing asphalt or bitumen
- Commercial cleaning products other than those specified
- Adhesives, paints, sealers, stains, and other finishes except as specified
- Any building materials contaminated by mold or mildew
- Any building materials or components that have been contaminated while in storage or during shipment

Contact the architect if a substitute product has not been specified for any application where the above substances would normally be used.

Product Substitution Procedure

Contractors will often ask to substitute a product different from the one that you have specified. The specified product may be unavailable, too expensive, too difficult to apply, or contractors may have one that they have used before and prefer. New and healthier products continue to be developed; it may be worth your while to consider certain substitutions. The first step in researching alternatives is to examine the Materials Safety Data Sheet (MSDS). You may also request a physical sample. In order to ensure that no substitutions are made without your consent or that of your architect, you may wish to add the following language to your agreement.

Product Substitutions

No products may be substituted for the specified product unless agreed upon in writing by the owner or architect. An MSDS must be provided on any substitution in order for it to be considered. Submit a physical sample to the owner or architect whenever possible.

Material Safety Data Sheets (MSDSs)

The Material Safety Data Sheet (MSDS) provides information about the chemical substances in a product, its handling precautions, and known health effects. The responsibility for preparing the MSDS lies with the chemical manufacturer. All companies are required to have an MSDS for every hazardous chemical they use. The information that must be included is listed below.

- With the exception of trade secrets, the specific chemical name and common names for hazardous ingredients
- Physical and chemical characteristics
- Physical hazards
- Health hazards
- Primary routes of entry to the body
- OSHA Permissible Exposure Limit (PEL) and any other recommended exposure limit
- Whether the chemical is a confirmed or potential carcinogen
- Precautions for safe handling and use
- Emergency and first aid procedures
- Name, address, and telephone number of manufacturer or other responsible party

MSDSs can be obtained from either the distributor or the manufacturer of the product in question.

What An MSDS Will Not Tell You

There is important information an MSDS does not reveal. Due to the Trade Secrets Act, companies are not required to list ingredients that they define as trade secrets. The OSHA Hazard Communication Standard requires that an MSDS list all health effects, yet health effects of trade secret ingredients can be exempted.[1]

The consumer is not allowed access to this information. However, one of the codes under the same law (OSHA Hazard Communication Standard 29, Code of Federal Regulation 1910.1200) permits physicians and other health care providers to access all product ingredient information for diagnostic and treatment purposes. Most doctors are unaware of their right to know.

Another significant omission regards the lack of disclosure of the "inert" ingredients which can account for up to 99% of product volume. Some of these so-called inert ingredients are more hazardous than the active ingredient(s).[2]

The permissible exposure levels (PELs) set by OSHA and the threshold limit values (TLVs) established by the American Conference of Governmental Industrial Hygienists (ACGIH) are misleading. Industry interests have played a major role in establishing these exposure limits. Most of these levels were established without prior testing.[3]

The small amount of testing that has been carried out was based on exposing rats to a single dose of a single chemical with cancer or death as the end point. In reality people are exposed to hundreds of chemicals at a time, the effects of which can be synergistic and accumulate in the body tissues over time. Monitoring for cancer or death does not take into account the many noncarcinogenic effects of chemicals, such as damage to the nervous, endocrine, and immune system. It is important to recognize that workplace standards are not set according to the safety of the worker, but rather according to what is considered feasible for industry.

Health effects listed in the MSDS are often vague and misleading. They are most accurate when listing the acute, short-term effects of chemicals such as eye and nose irritation, rashes, and asthma. The data on chronic, long-term exposure is often lacking and does not take into account cumulative effects over time as well as synergistic effects with other chemicals.

How An MSDS Can Be a Useful Tool

Although the MSDS has shortcomings, it is still an important tool for people involved in construction. If you are not working with a physician/architect team knowledgeable about chemicals, the MSDS can be confusing to interpret. Certain rules of thumb can be used to evaluate a chemical listed in the MSDS. For example, if no special precautions are required when using the chemical, there are no known health effects, and cleanup involves only water, then you might assume that the chemical in question has relatively low toxicity. On the other hand, if it is recommended that you wear gloves and use a respirator in a well-ventilated area, the product is likely toxic. Certain chemicals should pique your concern, such as chlorinated or fluorinated compounds, and chemicals which contain toxins such as toluene, phenol, benzene, xylene, styrene, formaldehyde, and the heavy metals, to name just a few.

In summary, although you cannot base your decisions solely on information from the MSDS, it is nevertheless useful. Appearing below are two MSDS examples. Product and manufacturer names have been omitted. Because MSDSs do not always follow a consistent format, comparisons can be difficult. Section numbers will vary but the information covered remains the same. While one example is indicative of a product that may be safe to use and in fact is a product that we recommend to our clients, the other is extracted from an MSDS that provides cause for concern.

Product Identification

This section includes the name of the product, the manufacturer, the date the MSDS was prepared, and by whom. In the first sample MSDS, the product is a wood preservative. The second sample involves a foam insulation material. As suggested in the examples, product identity information may vary, from nothing to substantial.

Example 1

Material Safety Data Sheet	
Section I - Product Identity	Manufacturer's Name
Data Prepared	
Preparer's Name	
Chemical Name: Water based wood preservative	Product
Chemical Formula: N/A (Product is a mixture)	Product Identification No.
DOT Shipping Class: Not regulated	
Emergency Telephone No.	

Example 2

Material Safety Data Sheet	
Manufacturer:	Date Prepared
Telephone Numbers:	
Emergency Number:	
Technical Information:	
(Regular Business Hours)	

Material Identification and Hazardous Components

This section lists the chemical names of all ingredients in the product found to be reportable health hazards. Exposure limits in some instances are established by government agencies. OSHA PEL refers to the permissible exposure limits set by OSHA. ACGIH TLV refers to the threshold limit values set by the American Conference of Governmental Industrial Hygienists. These values are updated on an annual basis.

If you are not familiar with the toxicity of the chemicals listed and you have no references available on the subject, then you can infer this information by examining the limits set by the government. When the limit is in parts per million, then you can be sure that the product is highly toxic. NE stands for no established limit, and could mean either that adequate testing has not been performed or that the product is not considered highly toxic.

Example 1

Section II - Hazardous Ingredients				
Hazardous Components (Special Chemical Identity/Common Name(s)*	CAS #	Wt. %	OSHA PEL	ACGIH TLV
Propylene Glycol	57-55-6	30-50	None established	None established
Polethylene Glycol	25322-68-3	30-50	None established	None established
Disodium Octoborate Tetrahydrate	12008-91-2	20-30	15 mg/m^3 (dust)	10 mg/m^3 (dust)
* Denotes a toxic chemical reportable under SARA Title 111 Section 313, Supplier Notification provision				
HMIS Information: Health: 1 Flammability: 0 Reactivity: 0				

With a health rating of 1, and 0 level flammability, and reactivity, and no established exposure limits, we can assume that the ingredients in this product are relatively safe.

Example 2

Section II - Hazardous Ingredients/Identity Information				
CHEMICAL NAME	CAS NO.	OSHA PEL	ACGIH TLV	PERCENTAGE
Polyurethane Resin	NE*	NE*	NE*	50-85
4,4-Diphenylmethane Diisocynate	101-68-8	0.02ppm CEIL	0.005ppm TWA	5-15
Chlorodifluoromethane (HCFC-22)	75-45-8	1,000ppm TWA	1,000ppm TWA	15-25
* Not established				
Hazard Rating: Health 3 Flammability 0 Reactivity 1				

In the second sample MSDS, the chemicals 4,4-diphenylmethane diisocyanate and chlorodifluoromethane (HCFC-22) are limited to parts per million. Both chemicals are in fact known to be extremely toxic. Diisocyanates and halogenated hydrocarbons can damage the nervous, immune, and endocrime systems with prolonged or repeated exposure.

Physical and Chemical Characteristics

This section describes how the material behaves. The information is useful for the design of ventilation systems and for providing adequate equipment and procedures for fire and spill containment.

Vapor pressure tells you how much vapor the material may give off. A high vapor pressure indicates that a liquid will easily evaporate.

Vapor density refers to the weight of the pure gaseous form of the material in relation to air. The weight of a given volume of a vapor (with no air present) is compared with the weight of an equal volume of air.

Specific gravity tells you how heavy the material is compared to water and whether it will float or sink.

Evaporation rate refers to the rate at which a material changes from a liquid or solid state to its gaseous form.

Volatile organic compounds (VOCs) provide you with an idea of the degree to which the substance will outgas. If the material is toxic, the degree of volatility would be an important point to consider.

Water reactive indicates whether the chemical reacts with water to release a gas that is flammable or presents a health hazard.

Appearance and odor indicate how a product is supposed to look and smell. For example, if the product is supposed to be clear and odorless but arrives on site with an acrid smell and/or appears cloudy, the product may be contaminated.

Example 1

Section III - Physical Characteristics			
Boiling Range	>369° F	Vapor Pressure (mm Hg.)	125mm Hg @ 100° F
Specific Gravity (H_2O=1)	1.1 - 1.3	Vapor Density (Air = 1)	> 1
% Volatile (Volume)	< 1%	Evaporation Rate (Bu)Ac = 1	> 1
Volatile Organic Content (VOC)	3.8 lb./gal.		
Solubility (specify solvents)	Miscible in water, alcohol, acetone, some glycol ethers, insoluble in petroleum hydrocarbons		
Appearance and Odor	Clear, odorless liquid		

NOTE: In the above MSDS, the evaporation rate is compared with the rate of evaporation for butyl acetate. With this particular product, the evaporation rate is less than butyl acetate.

Example 2

Section III - Physical/Chemical Characteristics		
Boiling Point	HCFC-22 Polyurethane Resin	-41.4° F at 1 ATM NE*
Vapor Pressure	HCFC-22	136 psia at 70° F
Vapor Density (AIR = 1)	HCFC-22	2.98 at 1 ATM
Specific Gravity (H_2O = 1)	Polyurethane Resin	1.1
Solubility in Water	Insoluble, reacts with water	
Appearance and Odor	Gel under pressure. Faint ether-like odor.	

NOTE: HCFCs are fluorinated carbons that are harmful to the ozone layer.

Fire and Explosion Hazard Data

The flash point tells you the minimum temperature at which a liquid will give off enough flammable vapor to ignite. Obviously, the more stable the product, the safer it will be.

Reactivity Hazard Data

This section can provide you with clues regarding the toxicity of a product.

Example 1

Section IV - Fire and Explosion Hazard Data			
Flash Point (Method Used)	Nonflammable	Flammable Limits (% in air)	Nonflammable
Extinguishing Media	Nonflammable		

Example 1

Special Fire Fighting Procedures Nonflammable	
Unusual Fire and Explosion Hazards None known	
Reactivity: Stable	Conditions to Avoid: Avoid extreme heat
Hazardous Polymerization: May Not Occur	Conditions to Avoid: None known
Incompatibility (Materials to Avoid) None known	
Hazardous Decomposition or By-products None known	

As seen in the Example 1 MSDS, the product is stable, with no incompatibility with other products, and without hazardous decomposition or by-products.

Example 2

Section IV - Fire and Explosion Hazard Data
Flash Point Polyurethane Resin >400° F
Extinguishing Media Water fog, foam, CO_2 or dry chemical
Fire Fighting Procedure: Wear self-contained breathing apparatus and turnout gear. Hazardous decomposition products include CO, CO_2, NO, and traces of HCI. Cured foam: Wear self-contained breathing apparatus. Hazardous decomposition products include CO, CO_2, NO, and traces of HCI.
Usual Hazards: Temperatures above 120° F will increase the pressure in the can, which may lead to rupturing. Cured foam: This product will burn. Do not expose to heat, sparks or open flame. This product is not intended for use in applications above 250° F (121° C). Always protect foam with approved facings. This product is not a FIRE STOP or FIRE BARRIER penetration sealant.
Section V - Reactivity Data
Stability: Stable under normal storage and handling conditions. Do not store above 120° F. Cured foam will deteriorate when exposed to UV light.
Incompatibility: Water, alcohols, strong bases, finely powdered metal such as aluminum, magnesium or zinc. and strong oxidizers.
Conditions/Hazards to Avoid: Contamination with water may form CO_2. Avoid high heat, i.e., flames, extremely hot metal surfaces, heating elements, combustion engines, etc. Do not store in auto or direct sunlight.

The Example 2 product is unstable when exposed to ultraviolet light and high heat, and is incompatible with many substances.

Health Hazard Data

This section provides useful information that will help you to determine the toxicity of the product in question.

Example 1

Section V - Health Hazard Data
Route(s) of Entry: Eye contact, inhalation, ingestion
Acute Health Effects: EYE CONTACT: May cause redness or irritation
INHALATION: N/A In sufficient doses may cause gastrointestinal irritation
SKIN CONTACT: N/A
Chronic Health Effects: Not listed as a carcinogen by the NTP, IARC, or OSHA; no adverse long-term effects are known.
Medical Conditions Generally Aggravated by Exposure: No adverse long-term effects are known
Emergency & First Aid: Eye contact: Wash with clean water for at least 15 min. If irritation persists, get medical attention.
INHALATION: N/A
INGESTION: If irritation persists, get medical attention.
SKIN CONTACT: N/A

Examining the health hazard section in the Example 1 MSDS would provide reassurance. The product appears to be only an irritant, with no known long-term health effects.

Example 2

Section VI - Health Hazard Data		
Toxicology Test Data	MDI:	Rat, 4 hr inhalation LC50 - Aerosol 490 mg/m^3 - Highly Toxic
		Rat, 4 hr inhalation LC 50 - Vapor 11 mg/l - Toxic
		Rat, oral LD 50 - > 10,000 mg/kg - Practically Nontoxic
		Rat, inhalation oncogenicity study - @ ~0.2, 1, 6 mg/m^3; URT irritant; Carcinogenic @ 6 mg/m^3
	HCFC-22:	Rat, 2 hr inhalation LC50 - 200,000ppm
Acute Overexposure Effects: Eye contact with MDI may result in conjunctival irritation and mild corneal opacity. Skin contact may result in dermatitis, either irritative or allergic. Inhalation of MDI vapors may cause irritation of the mucous membranes of the nose, throat, or trachea, breathlessness, chest discomfort, difficult breathing and reduced pulmonary function. Airborne overexposure well above the PEL may result additionally in eye irritation, headache, chemical bronchitis, asthma-like findings or pulmonary edema. Isocyanates have also been reported to cause hypersensitivity pneumonitis, which is characterized by flu-like symptoms, the onset of which may be delayed. Gastrointestinal symptoms include nausea, vomiting and abdominal pain. HCFC-22 vapor is irritating to eyes. Liquid is irritating to eyes and may cause tissues to freeze. Contact of liquid with skin may cause tissue to freeze (frost bite). Dense vapor displaces breathing air in confined or unventilated areas. Inhaling concentrated vapors can cause drowsiness, unconsciousness, respiratory depression and death due to asphyxiation. This compound also increases the sensitivity of the heart to adrenalin, possibly resulting in rapid heartbeat (tachycardia), irregular heartbeat (cardiac arrhythmias), and depression of cardiac function. Persons with preexisting heart disease may be at increased risk from exposure. Polyurethane resin forms a quick bond with skin. Cured foam is hard to remove from skin. May cause eye damage.		

Example 2

Chronic Overexposure Effects: Acute or chronic overexposure to isocyanates may cause sensitization in some individuals, resulting in allergic symptoms of the lower respiratory tract (asthma-like), including wheezing, shortness of breath and difficulty breathing. Subsequent reactions may occur at or substantially below the PEL and TLV. Asthma caused by isocyanates, including MDI, may persist in some individuals after removal from exposure and may be irreversible. Some isocyanate sensitized persons may experience asthma reactions upon exposure to nonisocyanate containing dusts or irritants. Cross sensitization to different isocyanates may occur. Long-term overexposure to isocyanates has also been reported to cause lung damage, including reduced lung function, which may be permanent. An animal study indicated that MDI may induce respiratory hypersensitivity following dermal exposure.
Carcinogenicity: Results from a lifetime inhalation study in rats indicate that MDI aerosol was carcinogenic at 6 mg/m^3 , the highest dose tested. This is well above the recommended TLV of 5 ppb (0.05 mg/m^3). Only irritation was noted at the lower concentration of 0.2 and 1 mg/m^3 . Lifetime exposure of rats to 5% HCFC-22 in air resulted in a slightly higher incidence of fibrosarcomas (a malignant connective tissue tumor) in male rats compared to controls. Some of these tumors involved the salivary glands. This effect was not seen in female rats at the same dose level or in rats of either sex at the lower dose level of 1%. Rats given HCFC-22 orally also showed no increased incidence of tumors. In addition, mice exposed to 5 and 1% HCFC-22 in a similar fashion showed no increased incidence of tumors. Spontaneously occurring fibrosarcomas are not uncommon in aging rats and the increase seen in male rats may have been due to a weak tumor promoting effect or other nonspecific effect (stress, etc.) of HCFC-22.
Mutagenicity: HCFC-22 has been shown to cause mutations in the bacterium salmonella. This may be due to the unusual metabolic capabilities of this organism. HCFC-22 is not mutagenic in yeast cell, hamster cell, or in vivo mouse and rat cell assays (dominant lethal and bone marrow cytogenic toxicity tests).
Teratogenicity: Offspring born to rats exposed to 5% of HCFC-22 for six hours per day during pregnancy showed stunted growth and a small, but statistically significant, incidence of absent eyes. However, this dose level also caused maternal toxicity. An increased incidence of absent eyes did not occur in rabbits exposed at 5% of HCFC-22 and below or in rats at 1% of HCFC-22 and below where maternal toxicity was not observed.
Medical Conditions Generally Aggravated by Exposure: Breathing difficulties, chest discomfort, headache, eye and nose membrane irritation.
Emergency and First Aid Procedures: Inhalation - Remove to fresh air. Give oxygen. If not breathing, give artificial respiration. Keep victim quiet. Do not give stimulants. Get immediate medical attention. Skin - If frostbitten, warm skin slowly with water; otherwise, wash affected areas with soap and water. Remove contaminated clothing and launder before reuse. Remove wet foam immediately from skin with acetone or nail polish remover. Dried foam is hard to remove from skin. If foam dries on skin, apply generous amounts of petroleum jelly or lanolin, leave on for one hour, wash thoroughly, and repeat process until foam is removed. Do not attempt to remove dried foam with solvents. Eye - In case of eye contact, flush with water for 15 minutes. Get immediate medical attention. Ingestion - In case of ingestion, get immediate medical attention.

In contrast, the Example 2 MSDS is not at all reassuring. This product is known to be carcinogenic, mutagenic, and teratogenic. It also may cause irreversible asthma, allergies, and other damage to the immune system. Although this product volatilizes quickly, the workers who install the product are exposed to an extreme health hazard.

Safe Handling Precautions and Leak Procedures

This section offers more clues regarding the safety of the product. The fewer the precautions given, the more reassuring the information.

Example 1

Section VI - Spill or Leak Procedures
Should be taken in case material is released or spilled: Soak up spill with absorbent material.
Waste Disposal Method: Dispose of in accordance with all local, state and federal regulations.

Example 2

Section VII - Precautions for Safe Handling and Use
Allow form to cure (harden).
Waste Disposal - Dispose according to federal, state, and local regulations.
Container Disposal - Dispose according to federal, state, and local regulations.
Storage - Store in cool, dry place. Ideal storage temperature is 60° F to 80° F. Storage above 90° F will shorten the shelf life. Do not store above 120° F (49° C). Protect containers from physical abuse. Do not store in auto or in direct sunlight. Store upright.

Control and Preventive Measures

This section lists the personal protective equipment that must be used, the type of ventilation to be used, and precautions to be taken when using the material for its intended purpose.

Example 1

Section VII - Special Protection Data
Respiratory Protection: None normally required
Ventilation: None normally required
Protective Gloves: None normally required
Other Protective Clothing or Equipment: None normally required
Section VIII - Storage and Handling Data
Precautions to be taken in handling and usage: Store in original container; keep tightly closed. Do not reuse container for other purposes. KEEP OUT OF REACH OF CHILDREN.
Other precautions: Read and observe all precautions on product label.

In the Example 1 MSDS, the product requires no special protective clothing or equipment, which is an indication of product safety.

Example 2

Section VIII - Personal Protection
Respiratory protection: None required if in well-ventilated area.
Clothing: Wear gloves, coveralls, long sleeve shirt, and head covering to avoid skin contact. Contaminated equipment or clothing should be cleaned after each use or disposed of.
Eye protection: Wear face shield, goggles, or safety glasses.
Ventilation: If ventilation is not enough to maintain P.E.L. exhaust area.

In the Example 2 MSDS, good ventilation and protective clothing over the entire body, including a face shield or goggles, are necessary.

The above MSDS examples demonstrate that the information supplied in the MSDS, although incomplete, is nevertheless useful. An MSDS allows you to obtain a general impression about the level of toxicity of many products you may consider using in home construction.

General Cleanup

Fortunately, inexpensive and effective alternatives to synthetic cleaning compounds are available. In fact, you can make several of them yourself. With homemade cleaning products, you do not have to buy a different product for every cleaning need. In this way, you can reduce cost, overall toxic exposure, and the amount of waste generated from packaging. Appearing below are some of the products that may be safely specified.

Household cleaning products are among the most toxic substances we encounter on a daily basis. It is ironic that our efforts to clean up often spread further contamination by spreading noxious fumes throughout the house. Moreover, these products end up down the drain where they pollute air, soil, and water.

Most commercial cleaning products are made from synthetic chemicals derived from crude oil. Labeling laws and the Trade Secret Act make it difficult to know exactly what is in any particular product. The product may contain highly toxic substances, but consumers have no way of knowing.

Some of the harmful ingredients found in commercial cleaning products include phenol, toluene, naphthalene, pentachlorophenol, xylene, trichloroethylene, formaldehyde, benzene, perchlorethylene, other petroleum distillates, chlorinated substances, ethanol, fluorescent brighteners, artificial dyes, detergents, aerosol propellants, and artificial fragrances.

Acceptable Cleaning Products

Brand Name Products

The following brand name cleaning products are acceptable:

- **AFM SafeChoice Safety Clean:** Industrial strength cleaner/degreaser and disinfectant
- **AFM SafeChoice Superclean:** All purpose cleaner and degreaser
- **AFM SafeChoice X158 Mildew Control:** Low odor, antifungal, antibacterial treatment
- **Bon Ami Polishing Cleanser:** Nonchloride, all purpose scouring powder
- **Mystical:** Odorless cleaner and deodorizer
- **Sodasan Soap Cleaner:** For cleaning floors and surfaces
- **SDA 1600:** Spectracidal disinfectant agent; benign germicide.

Common Household Products

The following common household products may also be used for cleaning:

- **Baking soda** cleans, deodorizes, scours, and softens water. It is noncorrosive and slightly abrasive, and is effective for light cleaning.
- **Borax** cleans, deodorizes, disinfects, and softens water. It is also effective for light cleaning, soiled laundry in the washing machine, and for preventing mold growth.
- **Hydrogen peroxide** (H_2O_2) is effective in removing mold. Purchase a 10% food grade solution. (The solution most commonly sold off the shelf is only 3%.) Use protective gloves to apply. A 10% solution will bleach many types of surfaces.
- **Soap** (as opposed to detergents) biodegrades safely and completely. It is an effective and gentle cleaner with many uses. For hands, dishes, laundry, and light cleaning, use the pure bar or soap flakes without perfume additives.
- **T.S.P.** (trisodium phosphate) as per manufacturer's instructions for grease removal. T.S.P. is available in hardware stores. Surfaces cleaned with T.S.P. should be neutralized with baking soda prior to the application of finishes. Note: Fluids containing T.S.P. should not be disposed of in septic systems or sewer systems due to high phosphate content.
- **Vodka** is effective for dissolving alcohol-soluble finishes.
- **Washing soda** (sodium carbonate) cuts grease, removes stains, disinfects, and softens water. It is effective for heavily soiled laundry and general cleaning purposes.
- **White vinegar** cuts grease and removes lime deposits. A safe and useful all purpose cleaning solution can be made from distilled white vinegar and plain water in a 50/50 ratio. For window cleaning, add 5 tablespoons of white vinegar to 2 cups of water. The solution should be placed in a glass spray bottle. Glass is preferred because plastics are known to release hormone disrupting chemicals into bottle contents. Vinegar has been used to clean and control mold growth, but the thin film of residue left on the surface supplies nutrients for new growth.

Case Study:

Toxic fumes from cleaning products

L.G. is a 53-year-old woman who was in reasonably good health until two years after she began working for a hotel as a housekeeper. At that time she consulted with Dr. Elliott complaining of rashes, headaches, joint pain, and fatigue. After extensive questioning, Dr. Elliott concluded that the source of her symptoms was probably found at her place of employment. Through a process of elimination, it became apparent that she had become sensitized to the pine scented product she used to disinfect bathrooms. Although this woman was unable to convince her employer to switch to less toxic cleaning products, her symptoms improved when she was transferred to a different job within the same building.

Discussion

Certain strong smelling cleaning products and disinfectants contain phenol which is known to sensitize the immune system in some people as happened to the unfortunate woman described in the case study above. When checking for phenol as an ingredient in a product, a generally safe rule of thumb is to look for any ingredient ending in "ol" or including phenol in its name.

Endnotes

1. Kersten, Elizabeth and Bruce Jennings, Pesticides and Regulation: The Myth of Safety. California Senate Office of Research, Senate Reprographics, April 1991 (41 pages).
2. Robert Abrams (Attorney General), "The Secret Hazards of Pesticides," New York Department of Law, June 1991 (5 pages).
3. B. I. Castleman and G.E. Ziem, "Corporate Influence on Threshold Limit Values." American Journal of Industrial Medicine 13 (1988): 531-559.

Resource List

Product	Description	Manufacturer/Distributor
AFM SafeChoice Safety Clean	Industrial strength biodegradable cleaner and degreaser for high moisture areas.	AFM (American Formulating and Manufacturing); 350 West Ash Street, Suite 700; San Diego, CA 92101-3404; (800)239-0321, (619)239-0321
AFM SafeChoice Superclean	All purpose, biodegradable cleaner/degreaser.	Same
AFM SafeChoice X158 Mildew Control	Low odor liquid surfactant coating for prophylactic use where mold and mildew are likely to appear.	Same
Bon Ami Polishing Cleanser	Kitchen and bath scouring cleanser without perfumes, dyes, chlorines or phosphates.	Faultless Starch/Bon Ami Company; Kansas City, MO 64101-1200. Available in grocery and health food stores.
Sodasan Soap Cleaner	Derived from plant chemistry for cleaning floors and other surfaces.	Eco Design/Natural Choice; 1365 Rufina Circle; Santa Fe, NM 87505; (800)621-2591, (505)438-3448
Mystical	Odorless cleaner and deodorizer.	The Non-Toxic Hot Line; 3441 Golden Rain Rd. #3; Walnut Creek, CA 94595; (800)968-9355 (orders only), (510)472-8868
SDA 1600	Spectracidal disinfectant agent; EPA approved nontoxic germicide.	Apthecure, Inc.; 13720 Midway Road, Suite 109; Dallas, Texas 75244; (800)969-6601

Further Reading

Wilson, Cynthia. *Chemical Exposure and Human Health.* McFarland and Company, Inc., 1993. A reference to 314 chemicals with a guide to symptoms. We use this handy guide to supplement information from the MSDS.

Ashford, Nicholas and Claudia Miller. *Chemical Exposures: Low Levels and High Stakes.* Van Nostrand Reinhold, 1991. A scientific discussion of the mechanisms underlying chemical sensitivities.

Dadd, Debra Lynn. *Nontoxic, Natural and Earthwise.* J.P. Tarcher, 1990. This book contains a good selection on alternatives to toxic cleaning products.

Lab Safety Supply, Inc. *Preparing, Understanding, and Using Material Safety Data Sheets.* This booklet can be obtained from Lab Safety Supply Inc. at (800)356-0783.

Bower, Lynn Marie. *The Healthy Household.* Healthy House Institute, 1995. This book contains a useful section on household cleansers.

Division 2 - Site Work

Site Selection

Long before construction begins, you will choose the appropriate site. Several measures are discussed in this division that can be integrated into construction to maintain or restore your site in the most healthful way.

When the ancient Romans selected a site for housing, they gave careful attention to the health giving qualities of the land. To test the potential home site, cattle were confined to graze in the area for a specific period of time, after which they were slaughtered and the innards examined. If the animals had unhealthy livers, the site was abandoned.

Although in the United States, the relationship between the natural geography of a site and human health is rarely considered, extensive studies have been conducted on this subject by baubiologists in Germany. Baubiology is the study of how buildings and the environment affect human health. While still in its infancy in the United States, "baubiology" has become a household word throughout much of northern Europe. According to baubiologists' studies, people experience an increased incidence of medical problems such as insomnia, cancer, and other immune disorders by living in areas where the Earth's natural electromagnetic fields are disturbed. It is especially important to avoid sleeping over these so-called "geopathic zones."

Few people have the specialized equipment designed to detect these zones, yet a simple hand-held, water filled compass will frequently suffice. The needle will deviate abruptly from magnetic north when passed over a strong natural field. A reputable dowser may also be able to detect variations in the Earth's electromagnetic currents with dowsing rods. Since there is no formal certification or licensing program for dowsers, finding an authentic one may be difficult. In many regions, well drillers use dowsers regularly and may be able to give you an appropriate recommendation for your area.

Unfortunately, the health consequences from the natural conditions of the site pale in magnitude when compared to potential hazards created by humans. Listed below are guidelines for choosing a site.

- Choose a location that is relatively unpolluted.
- Determine the direction that prevailing winds blow and how they change seasonally. Consider what is upwind from you. Avoid industrial areas, power plants, and other major pollution producers.
- Avoid proximity to high voltage power lines, microwave relay stations, and cellular phone and broadcast towers. In general, one-tenth mile from high voltage power lines and one-half mile from microwave cellular and broadcast towers are adequate. Many public utilities will provide free site measurements for background magnetic field levels. Ensure that measurements are taken at a time when power lines in the area are operating at peak load, or have the magnetic field calculated based on peak load projections. Utility companies should provide this information in writing.
- Avoid sites adjacent to parking lots and traffic corridors.
- Crest locations generally have better air quality and more air movement than valley sites.
- If you are considering a site in a populated area, then analyze your neighborhood in terms of present use and future development. How are nearby empty lots zoned? Do the neighbors use pesticides? Is there wood smoke from wood stoves and fireplaces in the winter?
- In rural areas, research the pesticide usage practices of nearby agricultural establishments.

Professional Assistance in Site Selection

You may require assistance in selecting your site, especially when remedies to suspected problems may be costly. Industrial toxins in the soil, poor percolation for installing a septic system, or unstable soils are examples of conditions that might be causes for concern. We recommend that you make your offer to purchase contingent on inspections by professional consultants. In this way, you can prevent being obligated to purchase a contaminated or unacceptable site. The following sections describe some of the more common consultant specialties.

Phase I Environmental Inspector

If you are considering a property with a past of industrial or agricultural use or underground fuel tanks, or if old buildings are suspected of containing lead or asbestos, then a Phase I environmental audit should be conducted to identify the risks. Remember that up-front costs are minor compared to hazardous waste cleanup costs that may far exceed property value.

Geotechnical Consultant

If you are concerned about the geological structure of the site, then a geotechnical engineer should be consulted. This person will be able to troubleshoot problems such as high water tables, unstable soils, earthquake faults, and sink holes. Engineering solutions can be devised for many of these problems so that the costs of development can be determined before purchase of the land.

Septic Engineer

In many locations an engineered septic system plan is required before a building permit will be issued. The septic engineer, who is often a geotechnical engineer as well, will study the land formation and perform percolation tests to determine how the sewage waste projected for your development can best be handled. In areas with limited percolation, steep slopes, or high water tables, the installation of a proper septic system could be costly or even impossible. If such conditions exist, it is best to be informed prior to purchasing the land.

Water Quality Specialist

In the event that a site lacks water, it is important to determine the cost of obtaining it. If the site is served by the local municipality, then the water company may be able to give you an estimate. If the site is rocky, you may need to excavate trenches by blasting the surrounding rock. If you must drill a well, the neighbors or the local well driller can inform you of the depth of surrounding wells in addition to information about water quality. If the site already contains a well, ensure that the submersible pump was manufactured after 1979, or that it is safe. If the oils in older pumps contain PCBs, they represent a serious health threat should a rupture occur.

Whatever the source of potable water, it should be tested by a professional and filtered or purified as required. Water quality will be discussed further in "Division 11 - Equipment."

Site Clearing

Although it is more convenient for a contractor to build on a site without obstacles such as trees, native vegetation, and boulders, many contractors will go to great lengths nevertheless to preserve as much natural vegetation and other landscape features as possible. However, do not assume that the preservation of your site will be a priority of the same magnitude for a contractor as it will be for you. To clarify your intentions and the contractual obligations of the contractor in this regard, you can specify these items. The following sample language could be included in the specification document.

Site Clearing

- It is the owner's intention to preserve the natural vegetation and land features of the site to the greatest extent possible.
- The owner and architect will approve the site layout prior to digging the footings.
- Topsoil and large boulders will be stockpiled for future use by the owner.
- All trees designated for removal from the building site are to be marked for review by the owner or architect.
- All stumps will be removed and disposed of off site in order to prevent insect infestation.
- The owner and architect will determine which trees are to be transplanted or maintained during construction. The contractor will provide the price for this service which is not included in this contract.
- The construction area and access to the construction area will be as small an area as is reasonable to facilitate construction of the home. This area is to be clearly demarcated and roped off to prevent any destruction of natural terrain outside the area by construction vehicles.

Grading

Many mold problems have been caused by poor drainage around the building perimeter which can cause water to puddle against the building and sometimes to seep to the inside. Although less prevalent in dry climates, mold is still a serious health threat. In flat roof construction where canales are used for roof drainage, erosion around canales is common even when the initial grading was adequate. The following specification is recommended.

Final Grading

- Water must have positive drainage away from the building at all points along its perimeter.
- All canales and downspouts will have splash blocks and an adequate drainage path away from the building.

Soil Treatment

Sometimes the soil under brick walkways, interior brick pavers, or the structure itself is treated with insecticides or herbicides. This practice should be avoided because many people have become sensitive to very low levels of pesticide exposure. Children are especially vulnerable. Some harmful agents will remain potent long after the building is gone. If there is cause to treat soil against possible insect or weed invasion, the use of boric acid, diatomaceous soil, or other non-toxic measures is recommended. (Refer to section on integrated pest management in "Division 10 - Specialties.")

Soil Treatment

- Do not treat soil with manufactured chemical treatments.
- Treat sand surfaces under floors and brick or stone walkways with diatomaceous soil.
- Use a barrier cloth under exterior walkways to prevent weed overgrowth.

Pavement

Petroleum tar, which is the main component of asphalt or "black top" paving, is carcinogenic and should be avoided. Not only does it emit harmful vapors during installation, but it will also volatilize when heated by the sun. More healthful options include concrete slab, concrete or brick pavers, and paving stone or gravel over a well-drained and compacted base.

Resources and Further Reading

International Institute for Bau-Biologie and Ecology. *Home Study Courses and Seminars.* P.O. Box 387, Clearwater, FL 34615; (813)461-4371.

Division 3 - Concrete

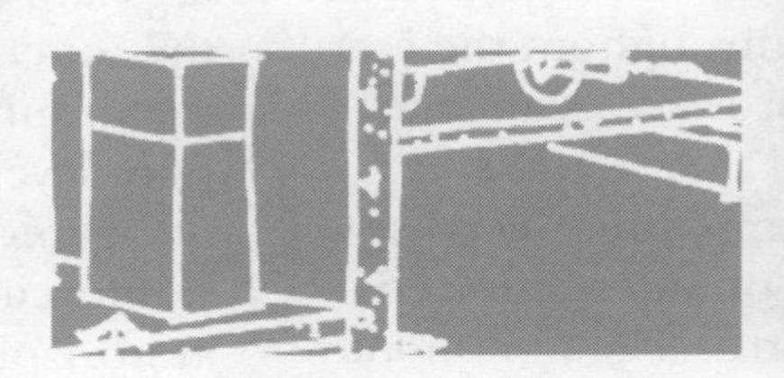

Concrete and concrete block are widely used in residential construction for footings, stem walls, exposed basement flooring, as a subfloor for slab on grade construction, and as a finished floor material.

Concrete consists of cement (usually Portland cement), aggregate, and water. Concrete is high in embodied energy due to the tremendous heat required to make cement. Once cured, concrete becomes an inert product and is not usually associated with health problems. However, certain practices can make concrete harmful to human health and should be avoided. Some of these practices are discussed in the following sections.

Practices To Be Avoided

Contaminated Aggregate

The aggregate component accounts for 60 to 80 percent of concrete volume. Aggregate materials range in size from fine sand to crushed rock pieces. Sometimes recycled materials are used as aggregate and these may be a source of contamination. Recycled industrial waste products such as fly ash may contain hydrocarbons and sulfur. Other recycled materials such as crushed brick are highly absorbent and may have been exposed to atmospheric pollutants prior to being used in concrete.

Contaminated Water

Only clean potable water should be used.

Toxic Admixtures

Many different types of admixtures may be added to the concrete mix to modify various properties. For example, air entrainment admixtures disperse air bubbles throughout the concrete to improve resistance to freezing and thawing. Water reducing admixtures decrease the amount of water required. Retarders and accelerators modify the setting time of concrete.

Super plasticizers allow for lower water to cement ratios. They frequently contain sulfonated melamine, formaldehyde condensates, sulfonated napthalene, and other potentially harmful ingredients.

Water reducing agents and air entrainment admixtures are frequently added to concrete mixtures even when not specifically requested. Generally the exact ingredients of an admixture are proprietary information. Admixtures can be completely eliminated if concrete work is scheduled for warm weather and if the concrete supplier is aware of this requirement.

Steel Reinforcement

Metal throughout a structure may contribute to electro-pollution. Placing a mesh of welded wire fabric within a concrete slab to help prevent cracking is common. This practice has been shown to distribute unwanted voltage throughout the home. Several types of approved nonmetallic reinforcing fibers are now readily available to do the same job at very little increase in cost.

Steel bars are placed in footings and stem walls to reinforce the concrete. In situations where the ground is higher than the floor level, such as in a basement or behind a retaining wall, steel reinforcing is also present, often at horizontal and vertical intervals of 12 inches or less.

In cases where an owner wishes to eliminate large amounts of conductive metal from structures, it is possible to use fiberglass reinforcing bars. These rebars were originally designed for bridge construction because they do not rust, corrode, or dissolve from galvanic action. Because there are very few manufacturers of this expensive product, and because they cannot be bent on site and must be preordered, they have rarely been used in residential construction. If you plan to use fiberglass rebar, verify with local code enforcement officials whether they will accept a particular product prior to purchase. We have used fiberglass rebar successfully on several homes.

Reduction of electric fields can also be achieved by grounding metal rebar, but the presence of the metal can still provide a potential pathway for stray magnetic fields.

Toxic Form Coating

Formwork is usually coated with release agents so that it can be easily removed and reused once concrete has cured. Although many inert products may be used for this purpose, diesel fuel and other equally noxious substances are commonly used because they are less expensive.

Toxic Waste as Fuel

Toxic industrial waste products are sometimes burned in order to provide high temperatures required to make cement. We believe that these cement products should be boycotted, not necessarily because the cement is inferior, but because the use of toxic waste as fuel is an environmentally unfriendly practice. In addition, waste or fly ash is sometimes used as an aggregate in the concrete, resulting in a product with toxic properties.

Preventing Unacceptable Practices

Consider including the following specifications in order to avoid the abovementioned problems.

Basic Concrete Materials and Methods

Mix

- Concrete shall be made of Type I Portland cement, clean sand and gravel, and potable water.

Water

- Water shall be free of taste, color, and odor and should not foam or fizzle.

Aggregates

- Only clean, natural mineral aggregates are acceptable. The following are unacceptable aggregates: crushed brick, crushed sandstone, crushed concrete slag, fly ash, cinder, and volcanic material (other than pumice). The contractor shall verify the aggregate content with the concrete supplier prior to pouring.

Admixtures

- No admixtures, including accelerants or retardants, shall be used in the concrete. It is the contractor's responsibility to comply with the necessary climactic parameters so that required strengths and finishes are obtained without additives. Verify with supplier that all concrete is free of admixtures, including air entrainment and water reducing agents.

Concrete Forms and Accessories

- The use of petroleum based form oil as a release agent is prohibited.
- Use uncoated forms, or forms coated with vegetable oil or an acceptable paint, as specified in "Division 9 - Finishes."

Slab Reinforcing

- Slab reinforcing shall be 1/2" fiberglass or polypropylene fibers as manufactured by **Fibermesh** or **Fiber-Lock**. Check with local code officials to determine acceptability.
- Fiberglass rebar can be purchased from **Marshall Vega Corporation** or **Kodiak FRP Rebar**. Rebar must be properly grounded or fiberglass rebar is to be substituted for metal rebar.

Concrete Finishes

Colorants

- No aniline based coloring agents are to be used.
- Use only high quality mineral pigments such as **Chromix Admixture** and **Lithochrome Color Hardener** or **Davis Colors**. Verify with the manufacturer that the selected color is free of chromium and other heavy metals.

Slab Treatment

Concrete slabs can act as a wick for ground moisture, thereby promoting mold and fungal growth. A layer of gravel under the slab will break the capillary action. The slab can also be sealed to further prevent moisture from the ground from entering the building. Some sealers are solvent based and should be avoided. The following sealers are acceptable:

- **AFM Safecoat Paver Seal .003, AFM Safecoat MexeSeal, AFM Safecoat Penetrating Waterstop**—Water based sealers and finish coats.
- **Rub-R-Wall**—Asphalt free waterproofing membrane.
- **Sodium Silicate**—A clear sealer supplied by various manufacturers. Inhaling powdered sodium silicate must be avoided because it can cause silicosis.
- **Vocomp-25**—A solvent reduced, water based sealer.
- **Xypex**—Concrete waterproofing by crystallization.

Chart 3-1: Resource List

Product	Description	Manufacturer/Distributor
AFM Safecoat Paver Seal .003	A low odor, water based sealer for previously unsealed concrete that fills up pores and preps slab for top coat.	AFM (American Formulating and Manufacturing); 350 West Ash Street, Suite 700; San Diego, CA 92101-3404; (800)239-0321, (619)239-0321
AFM Safecoat MexeSeal	A satin finish top coat to be used over the Paver Seal .003.	Same
AFM Safecoat Penetrating Waterstop	A satin finish final coat that may be used over MexeSeal to further improve water repellence.	Same

Product	Description	Manufacturer/Distributor
Chromix Admixture	Mineral pigment containing no chromium or other heavy metals; for use in concrete.	L.M. Scofield Company; P.O. Box 1525; Los Angeles, CA 90040; (800)222-4100
Lithochrome Color Hardener	Mineral pigment containing no chromium or other heavy metals; for use in concrete.	Same
Davis Colors	Mineral based pigments for concrete.	Davis Colors; 3700 E. Olympic Blvd.; Los Angeles, CA 90023; (800)356-4848, (213)269-7311
Fiber-Lock	Polypropylene fiber additive reinforcement for concrete slabs.	Fiber-Lock Company; PO Box 1087; Keller, TX 76244; (800)852-8889, (817)498-0042
Fibermesh	Fiberglass reinforcing for concrete slabs.	Synthetic Industries; P.O. Box 22788; Chattanooga, TN 37422; (800)621-0444, (423)899-044
Kodiak FRP Rebar	Fiberglass reinforcing bars.	International Grating, Inc.; 7625 Parkhurst; Houston, TX 77028; (800)231-0115, (713) 633-8614
Lithochrome Color Hardener	See Chromix Admixture.	See Chromix Admixture source.
Nonmetallic rebar	Fiberglass reinforcing bars.	Marshall Vega Corporation; P.O. Drawer 400; Marshall, AK 72650; (501)448-3111
Rub-R-Wall	Rubber polymer waterproofing membrane containing no asphalt. Manufacturer claims product is nontoxic once dry.	Rubber Polymer Corporation; 1135 W. Portage Trail Extension; Akron, OH 44313; (800)860-7721
Sodium Silicate	Clear sealer for concrete floors. Widely distributed in hardware stores.	Ashland Chemical Inc.; 5200 Blazer Pkwy.; Dublin, OH 43017; (800)258-0711, (614)889-3333 Dupont Company; 1007 Market Street; Willmington, DE 19898; (800)441-7515
Vocomp-25	Water based acrylic concrete sealer.	W.R. Meadows; P.O. Box 543; Elgin, IL 60121; (800)342-5976, (708)683-4500
Xypex Concrete Waterproofing	EPA approved for concrete potable water containers. Protects concrete against spalling, effervescence, and other damage caused by weathering and bleeding of salt.	Xypex Chemical Corporation; 13731 Mayfield Pl.; Richmond BC Canada V6V2G9; (800)961-4477, (604)273-5265

Division 4 - Masonry or Other Alternatives to Frame Construction

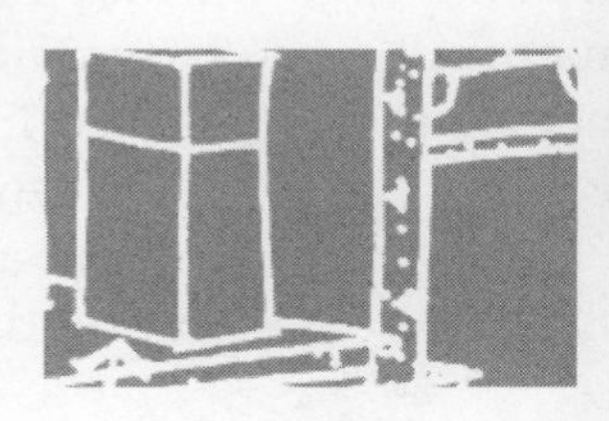

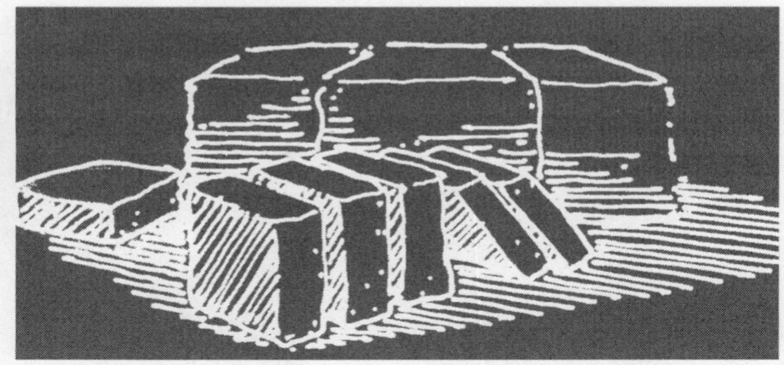

In recent years, with renewed interest in environmental concerns and energy efficiency, several pre-industrial methods of building have been revived which use natural, locally available materials such as adobe, straw bale, and rammed earth. A brief description of these methods is provided below. Publications dedicated to alternative construction methods are referenced at the end of this section.

Each of the abovementioned methods has great potential for healthy house building because the walls themselves provide insulation and they can be finished with a covering of plaster applied directly to the wall. The need for exterior sheathing, batt insulation, gypboard, joint fillers, and paint is eliminated. Many volatile organic compound (VOC) contamination sources are thereby eliminated as well. Because the walls are porous and interactive with the natural environment, temperature and moisture levels are modified and a slow exchange of air through the wall material takes place. This concept, known as a "breathing wall," is one of the basic principles of baubiology. Because all such methods use locally harvested and renewable resources, impact on the environment is minimal compared to methods based on manufactured materials.

Adobe

In the U.S. Southwest, adobe has been a traditional building material, predating building codes. Because the R-value of adobe is fairly low, it requires additional insulation in order to meet state energy requirements in all but the warmest portions of North America. (R-value, a measurement of thermal insulation, indicates resistance to heat flow. The U.S. Department of Energy has recommended R-values for every area of the United States. Higher R-values are recommended for colder climates.) A higher R-value is usually obtained by adding foam insulation to the exterior of the building which affects the "breathability" of the wall and marries an environmentallyfriendly product to an unfriendly one.

Adobe blocks are frequently "stabilized." One of the main reasons for this procedure is to prevent breakage during transport. The most common stabilizer is asphalt, a material that should be avoided in the healthy home. Unstabilized adobes can be purchased from some adobe yards. They can also be made on site with an adobe press, thereby obviating the need for stabilizers.

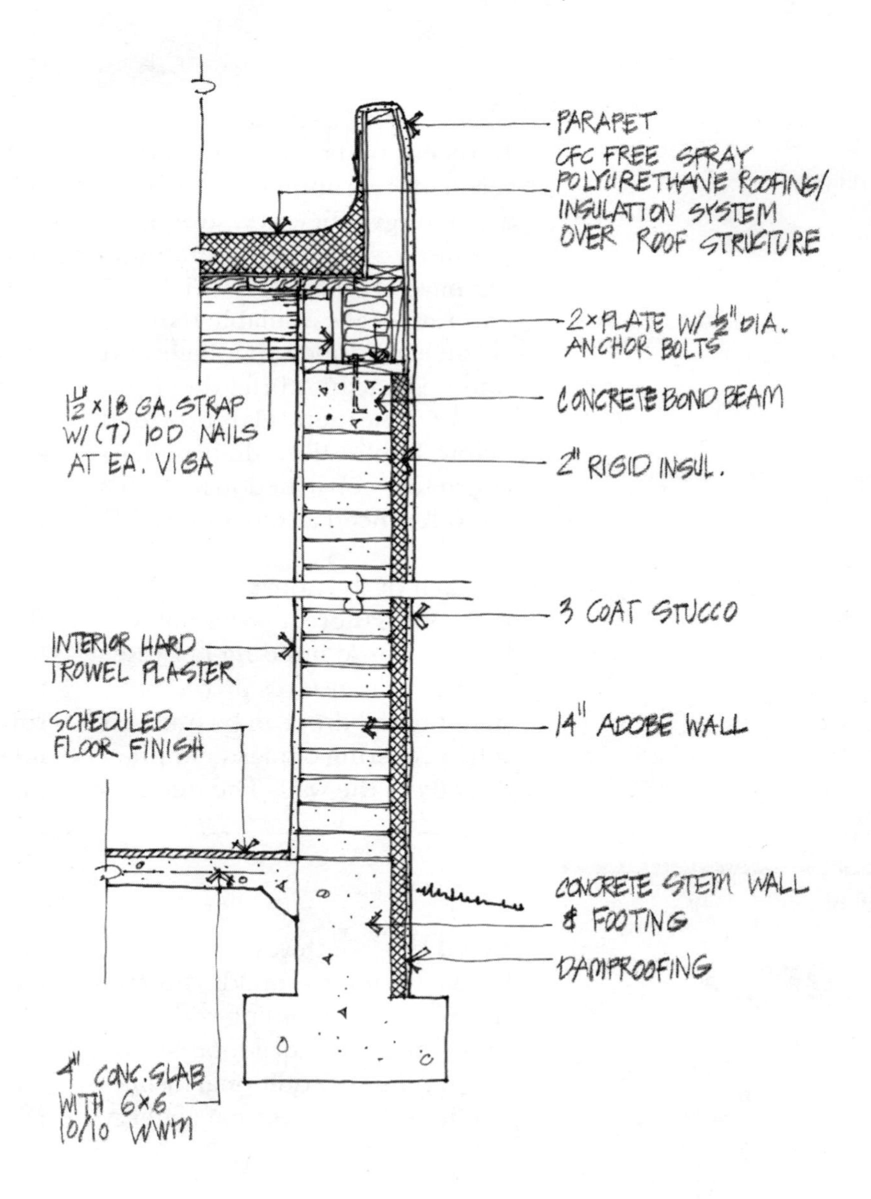

Cutaway of typical adobe wall.

Cob

Cob was a traditional form of building throughout preindustrial Europe. It has recently been revived in this country by The Cob Cottage Company. A mixture of moistened earth containing suitable clay and sand content is mixed with straw and formed into loaves which are then piled onto a wall and blended with the previous layers. The result is a monolithic, load bearing mud wall. At least 18" thick, the wall can serve as an insulating wall with no need for foam insulation and stabilizers.

The Cob method is time-consuming but does not require a high degree of skill or specialized tools. It lends itself to sculptural shapes.

Rammed Earth

Earth containing the proper moisture and clay content is rammed into formwork in 6 to 8" layers. The walls are thick, precise, and beautiful. Different colors of earth can be used to create decorative effects. The walls do not require plaster or further insulation. This technique is most suitable in arid climates. In earthquake zones, steel reinforcement may be required.

Straw Bale

A large amount of information is available about straw bale construction due to its current renaissance as a building material. Its thick, highly insulated walls are extremely energy efficient and beautiful. Several states including California, New Mexico, and Arizona issue building permits for straw bale construction. This new development results from the work of a few dedicated individuals who have spearheaded efforts to perform stringent fire and structural testing. These expensive tests are required for code approval. Whereas some states will permit load-bearing straw bale construction, others will permit it only as infill material between a post and beam structure.

Because much of the straw grown in the United States is heavily sprayed with pesticides, we recommend looking for straw that has been organically grown.

Bales of straw often contain mold. Because the walls are allowed to breathe, in theory the bales will always remain dry enough so that mold will not be a problem. Should water become trapped in the wall due to roof failure, plumbing leaks, poor drainage or other building systems failures, then ambient mold can become a problem. As with most building systems, water and moisture management strategies must be incorporated in the design in order to prevent mold growth.

Cutaway of typical straw bale wall

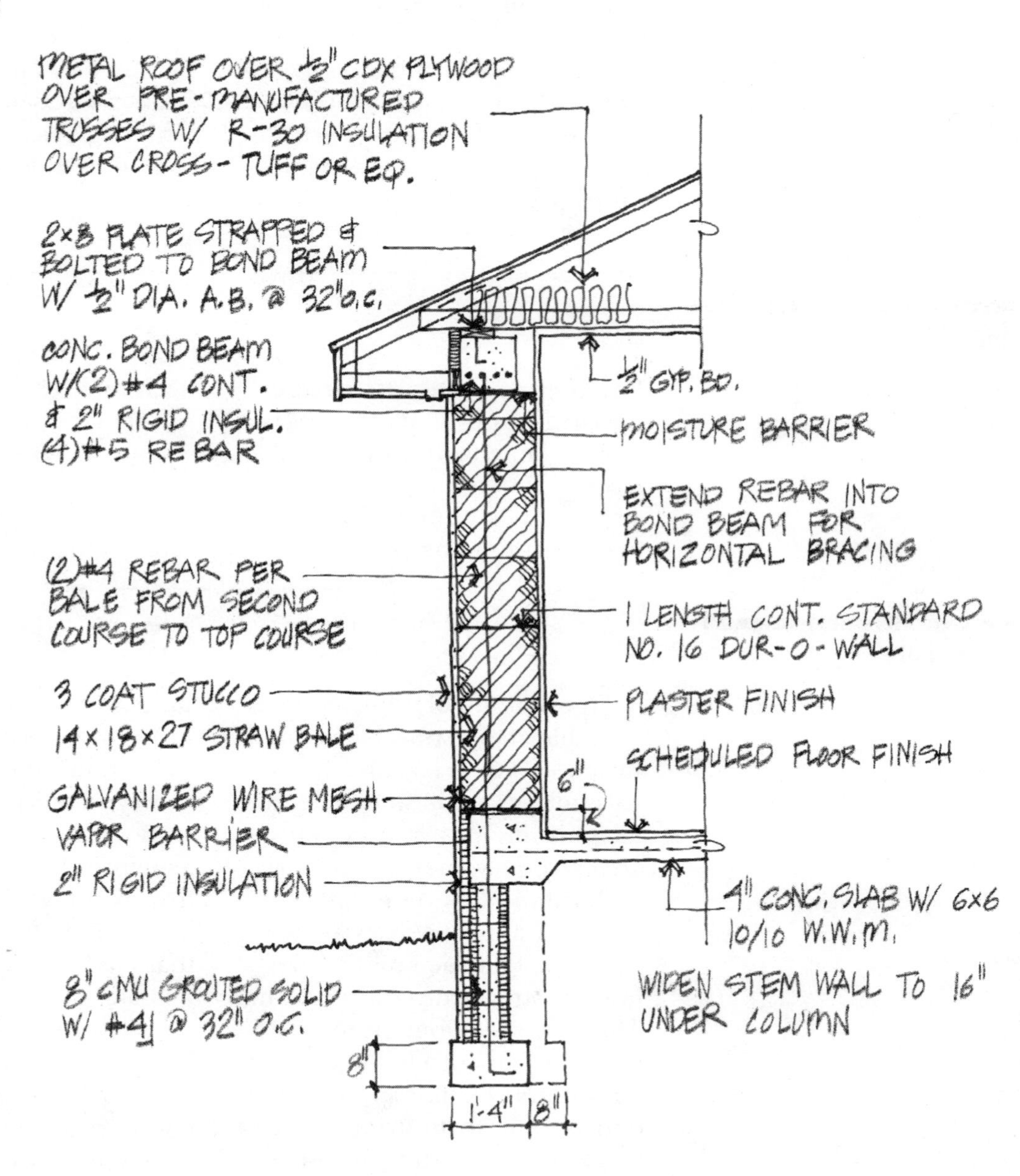

Straw/clay

Straw/clay construction combines the mass of adobe with the insulating capacity of straw. Straw is mixed with a clay soup and rammed into forms. This is an ancient European method of building that has only recently been introduced in this country by Robert Laporte, founder of the Natural House Building Center.

This technique uses straw clay as an "outsulating" wall around timberframe structures. The result is a precise wall that can accept mud plaster without any further wall preparation. The Natural House Building Center conducts workshops around the country to teach people how to build with straw clay.

Cutaway of typical straw/clay wall.

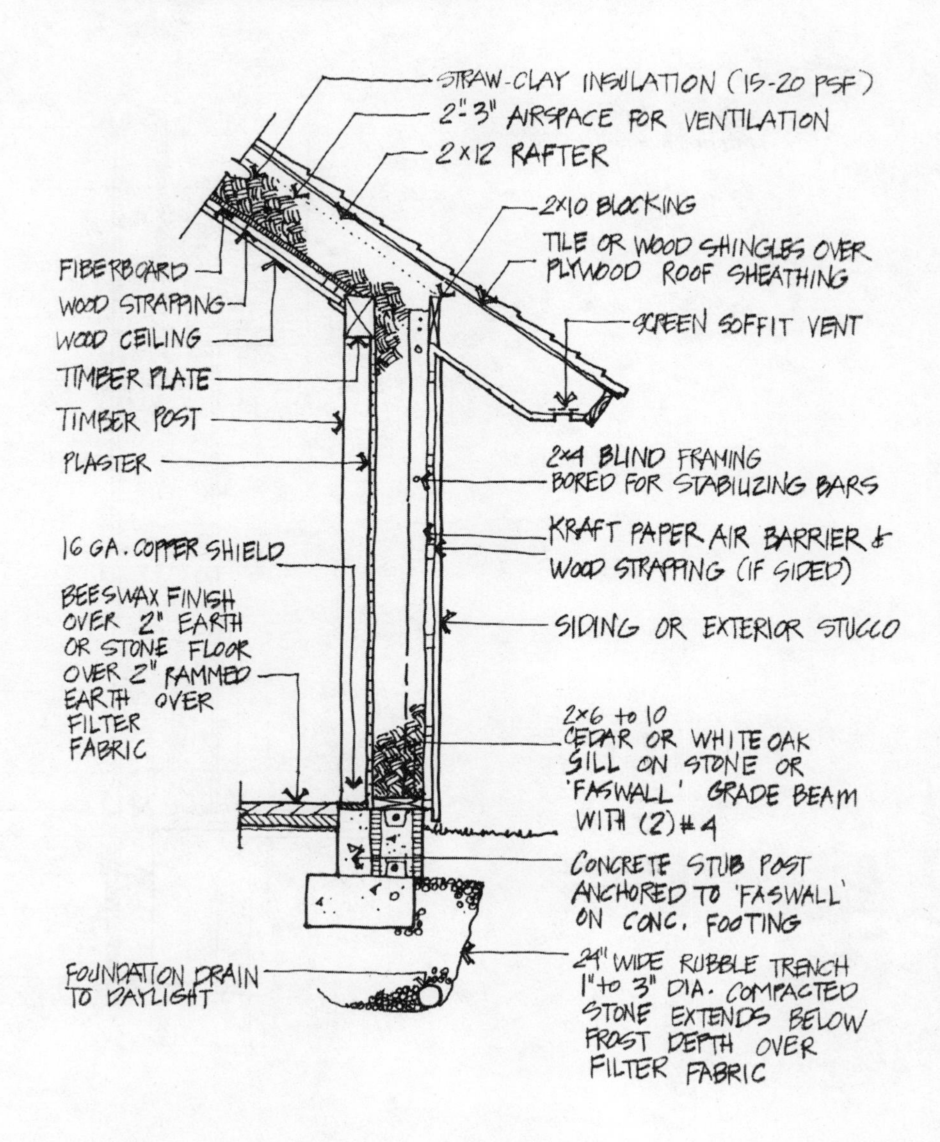

Pumicecrete

In this method, thick walls are created by mixing pumice, a very porous volcanic rock, with a light soupy concrete. The mixture is poured into formwork. The resulting walls have both thermal mass and a high insulation value, and are ready to accept plaster without further preparation. Since the wall uses a certain amount of concrete, the rules for concrete formwork and aggregate composition as outlined in "Division 3 - Concrete" must be followed. Pumice can be radioactive. Samples should be tested with a Geiger counter to be sure they are free of radioactive material. Refer to "Division13 - Special Construction" for testing methods.

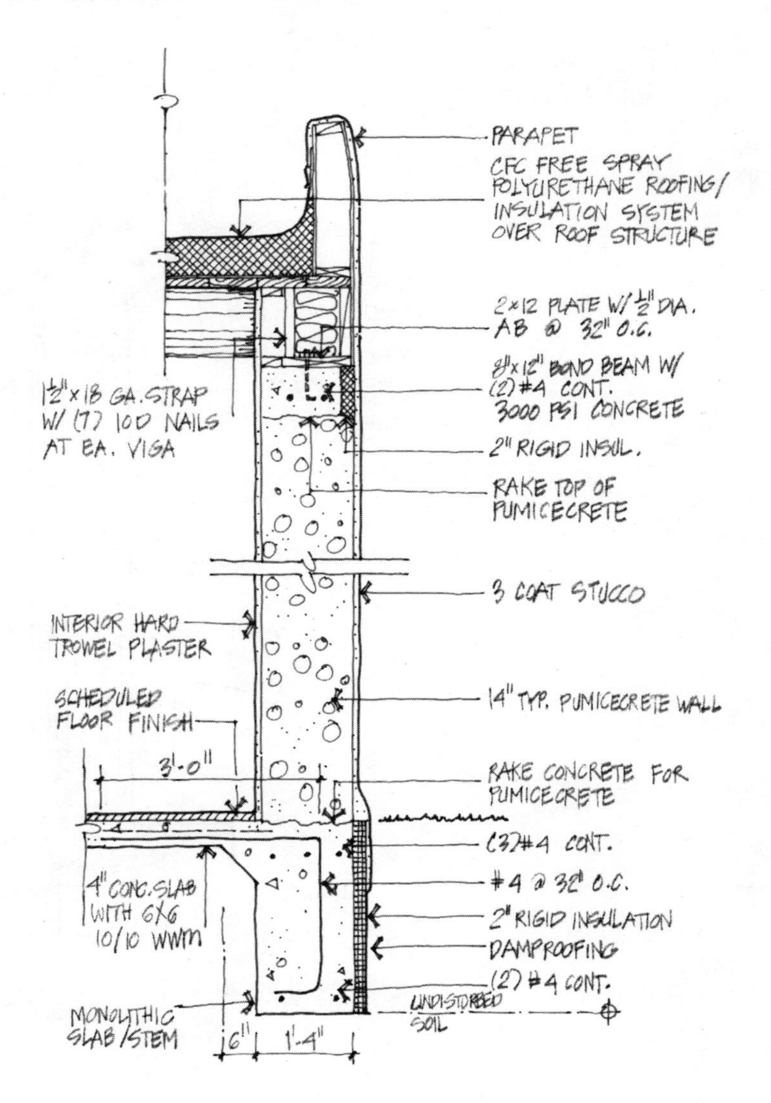

Cutaway of typical pumicecrete wall.

When planning to build with any of these materials, careful inquiry must be made into permit status with local building authorities. Regional factors such as drainage, rainfall, temperature, humidity, freeze and thaw cycles, and the availability of natural materials will make some of these solutions more suitable in certain locations than in others.

We would like to emphasize that the use of natural construction materials does not automatically create a healthy home because the building envelope is only one of many components. Sadly, we have seen many examples of healthy building construction in which the pristine indoor air was degraded by components introduced into the building envelope, such as toxic finishes or unhealthful heating systems. When natural construction materials can be used in conjunction with nontoxic interior components, it is possible to produce buildings of exceptional vitality.

Further Reading

Cob Cottage Company. *Earth Building and Cob Revival: A Reader, Third Edition.* Cottage Grove, OR: Cob Cottage Company, 1996.

Easton, David. *The Rammed Earth House.* White River Junction, VT: Chelsea Green Publishing Company, 1996.

Laporte, Robert. *Mooseprints: A Holistic Home Building Guide.* Santa Fe, NM: Natural Housebuilding Company, 1993. (505)471-8531.

MacDonald, S.O., and Matt Myhrman. *Build It with Bales: A Step-by-step Guide to Straw Bale Construction.* Treasure Chest Publications, 1997.

McHenry, Paul G. *Adobe: Build it Yourself.* University of Arizona Press, 1985.

Roodman, David Malin, and Nicholas K. Lenssen. *A Building Revolution: How Ecology and Health Concerns Are Transforming Construction.* Washington, DC: Worldwatch Paper 124, March 1995 (202/452-1999).

Steen, Athena Swentzell, Bill Steen and David Bainbridge. *The Straw Bale House.* White River Junction, VT: Chelsea Green Publishing, 1994.

Division 5 - Metals

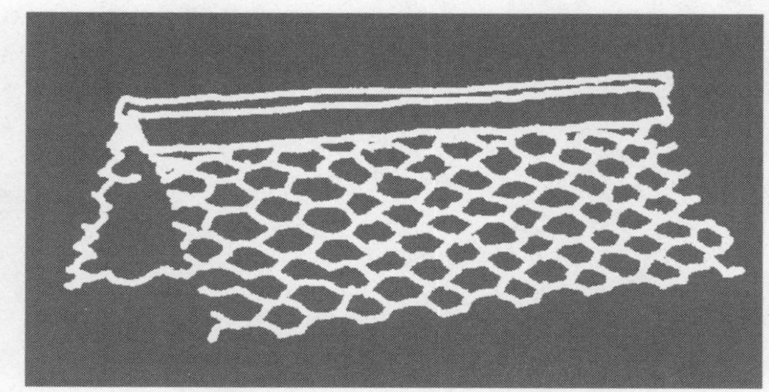

Oil Residue on Metals

Expanded metal lath and other metal goods are often shipped to sites coated in rancid oil residues left over from manufacture. Such residue will be odorous for a prolonged period of time unless the metal is cleaned. Oil removal is not standard practice. Consider adding the following to your specifications.

Oil Residue on Metals

- Remove oil residue from all coated metal products using a high pressure hose and one of the acceptable cleaning products listed in these specifications.

Tip: Some builders have found that the high pressure hoses at self service car washes are effective for removing oil residues.

Metals and Conductivity

The role that metals play in the electroclimate of a building, along with proper grounding considerations, will be discussed in "Division 16 - Electrical."

Metal Termite Shielding

Where floors are joisted, the proper application of metal termite shielding, as illustrated below, will create a physical barrier that is effective against subterranean termites.

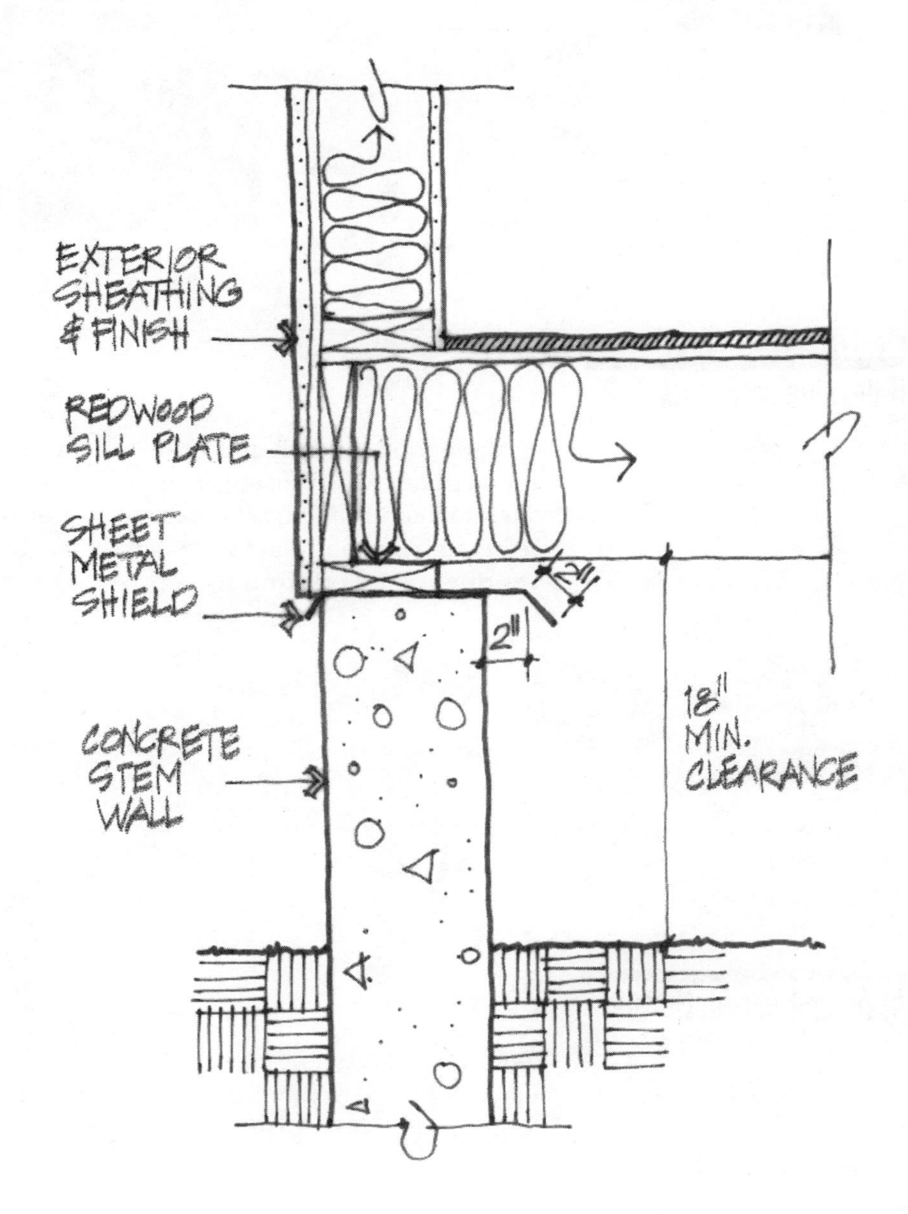

Termite shield detail.

Division 6 - Wood and Plastics

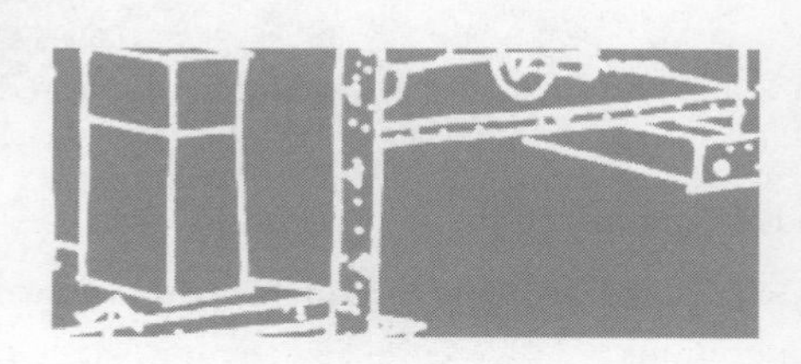

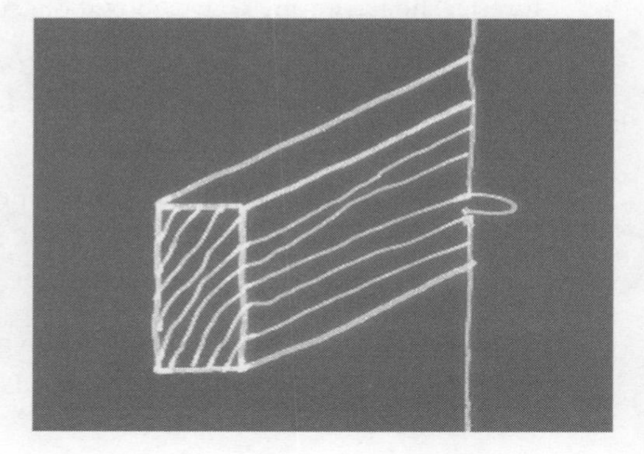

The history of forest mismanagement in the United States is a very long one, and beyond the scope of this book. Wood can be used in an ecologically conscious manner through sustainable harvesting and replanting, along with a commitment to a type of building that carries with it a longevity greater than the growth period of the trees from which the resulting structure is built.

Use of Sustainably Harvested Wood

Sustainably harvested wood can often be obtained for the same price as lumber harvested by environmentally damaging methods, such as clear cutting. By specifying the use of sustainably harvested woods for a building project, you are helping to raise awareness and increase market demand. The following language for wood specification has been created by the Smart Wood Program of the Rainforest Alliance and is reproduced here as sample language for your specifications. (See Chart 6-2, "Resource List," for sources of sustainably harvested wood.)

Use of Sustainably Harvested Wood

We encourage bids, proposals, or price quotations that include purchase of forest products from organizations whose products come from forestry operations which are formal participants in a credible, independent forestry certification program, such as the Rainforest Alliance's Smart Wood Program.

Participation of the forestry operations must have included, at a minimum, the following criteria:

- Been subject to a formal field level assessment
- Met minimum environmental, silvicultural, and social criteria
- Agreed to periodic and/or random audits of forestry operations, and
- Agreed to a specific timeline for continual improvement of forest management operations

Sources of Sustainably Harvested Wood

- **Quality Wood Products**: Harry Morrison owns this small sawmill operation in northern New Mexico. A former forester, he has strong convictions about sustainable harvesting and personally marks all trees to be felled in his operation.
- **Forest Trust Lumber Brokerage**: Brokers for lumber from small, independent New Mexico producers with sustainable harvesting practices.
- **Plaza Hardwood, Inc.**: The owners Paul and Toni Fuge specialize in sustainably harvested and recycled wood flooring.
- **EcoTimber International**: Quality hardwoods from environmentally sound sources.
- **Rainforest Alliance**: SmartWood Program is accredited to certify sustainably harvested wood operations. Publishes SmartWood List which identifies certified operations.

Health Concerns Associated with Wood Frame Construction

Wood has historically been used as a component of a "breathing wall" system, whether it be the half timber, waddle, and daub constructions of medieval Europe or the log cabins of our ancestors in North America. Wood is an advantageous material in a healthy home because it has the property of "hygroscoposity." This means that it has the ability to absorb and release moisture, thus balancing humidity levels and the electro-climate.

In standard home construction, the air space between the wood studs is filled with insulation laden with chemicals. The exterior sheathings often contain formaldehyde based glue or asphalt backing. The wallboard applied to the inside face of the studs is finished with harmful joint compounds. The studs sit on a sill plate that is pressure treated with a pesticide to prevent rot and insect infestation. When standard construction is the only option, we recommend the use of a barrier between the wall construction and living space. Barriers are only partially effective, however, and may trap moisture which promotes mold and rot. Furthermore, the potential health benefits of a "breathable" wall system are lost when a barrier is installed.

Some people have become sensitized to the terpenes in wood, and for them, exposed soft woods like pine, spruce, and cedar are intolerable and should be avoided. Hardwoods are not usually a problem.

Construction lumber is at risk of contamination by pesticides when farmed, and during transportation and storage. For those who have severe sensitivities, it is important to locate a source for uncontaminated wood. Woods that are sustainably harvested usually come from unsprayed environments. In some cases uncontaminated lumber can be picked up directly at the mill.

Wood Selection and Storage

Kiln dried framing lumber is drier than standard lumber. It is therefore more true to size and less susceptible to shrinkage and mold infestation. Certified, sustainably harvested, kiln dried framing lumber is now becoming widely available and can be obtained through Eco Timber International and the Forest Trust Lumber Brokerage. Framing lumber of this type is slightly more expensive than standard lumber which is often logged using unsustainable practices.

Wood may occasionally be delivered to the site containing mold or mildew. It can also become moldy while stacked on site if unprotected. The following instructions should be included in your specifications.

Wood Selection and Storage

- Framing lumber shall be kiln dried.
- Fir spruce and hemlock are preferred where available at no additional cost to owner.
- Only wood that is free of mold and mildew is acceptable.
- Wood stored on site shall be protected from moisture damage by elevating it off the ground and covering it with a tarp during precipitation.
- Wood that becomes wet must be quickly dried by cross stacking to promote aeration. It must have less than 19% moisture content, as tested by a moisture meter, and must be free of all signs of mold in order to be acceptable. See "Division 13 - Special Construction" for moisture meter testing.

Wood Treatment

Wood surfaces and edges exposed to the weather will usually be surface treated to make them more weather resistant, unless the wood has already been pressure treated or is naturally rot resistant. Pentachlorophenol and creosote are two commonly used wood preservatives that are quite toxic. Creosote is a dark colored, oily tar that will outgas harmful vapors long after it has been applied. Pentachlorophenol has been shown to cause fetal death and liver damage in adults and has been banned in some European countries. For less toxic wood protection you can consider using the following specifications.

Wood Treatment

The use of products containing creosote or pentachlorophenol is prohibited. The following products may be used for exterior wood preservation.

- **9400 W Impregnant**: Solvent free, water repellant, ultraviolet protective coating for wood interior/exterior
- **Bio Shield**: Wood preservative oils
- **Bora-Care**: Low toxicity, borate based, penetrating preservative used for protection against wood boring insects
- **Livos Donnos Wood Pitch Impregnation**: Penetrating preservative for wood that is in contact with moisture
- **Livos Dubno Primer Oil**: Undercoat for exterior wood
- **Old Growth**: A two-part process that provides antimicrobial and antifungal properties and imparts an aged patina to woods
- **OS Wood Protector**: Preserves against water damage, mold, mildew, and fungus
- **Polyshield**: Interior and exterior polyurethane wood protection
- **Shellguard** and **Guardian**: Borate based wood preservatives for protection against wood boring insects
- **Timbor**: Low toxicity, borate based wood preservative that protects against fungus, termites and wood boring beetles.
- **WEATHER PRO**: A water based, water repellant wood stain

Case Study:
Pesticide treated lumber

Although Germany has been a leader in the baubiology and healthy housing movement, it was only during the last decade that the general public became aware of multiple chemical sensitivity disorder. This awareness followed the experiences of a significant number of people who were exposed to lumber treated with both the preservative pentachlorophenol and the pesticide lindane. Hundreds of people developed chronic neurological complaints, chronic fatigue, and an unusually heightened sensitivity to chemicals that were previously tolerated. Lindane has subsequently been banned in Germany as a wood treatment.

Wood Adhesives

Wood adhesives commonly contain harmful solvents. However, solvent free solutions are readily available and may be specified.

The following adhesives are healthier choices for various wood related applications.

Wood Adhesives

- **100% Pure Silicone Caulk**: Can be used as a subfloor adhesive. Specify aquarium grade caulk without additives
- **AFM Safecoat Almighty Adhesive**: Low odor, nontoxic, water based adhesive for gluing wood and wood laminates
- **Elmer's Carpenter's Glue**: Low odor, nontoxic, water based glue for porous materials
- **Solvent Free Titebond Construction Adhesive**: For plywood, paneling, hardboard, and wet or frozen lumber
- **Timberline 2051 Wood Flooring Adhesive**: For laminated plank and parquet flooring
- **Titebond Solvent Free Subfloor Adhesive**

Rough Carpentry

Sill Plates

Sill plates or mudsills are decay and insect resistant wood members used in frame construction wherever wood comes into contact with concrete or soil. They are most commonly pressure treated so that the wood is impregnated with certain chemicals. The two most common chemicals used are chromated copper arsenate (CCA) and ammoniacal copper arsenate (ACA). CCA and ACA contain arsenic salts and chromium compounds that can leach out on the site and be absorbed through the skin or ingested by mouth. They are extremely toxic to both humans and the environment. We do not recommend them for the healthy home. CCA imparts a green tinge to the wood. You have probably observed this toxic wood being used in children's playground equipment.

For many centuries before these chemicals were formulated, builders had devised natural means for avoiding rot and insect infestation. They commonly charred the portions of wood that were to be placed in the ground or else used naturally resistant woods. Today, the standard building practice is to use pressure treated lumber. Pressure treated lumber will most likely be used in your home unless you ask for a substitute. For a healthier installation, consider adding the following specifications.

Sill Plates

- No wood treated with chromated copper arsenate (CCA) or ammoniacal copper arsenate (ACA) may be used on this job.
- Wood treated with ACQ (alkaline/copper/quat) is acceptable. (See ACQ Preserve in Chart 6-2, "Resource List.")
- The heartwood of untreated cedar or redwood is acceptable for use as sill plates. Obtain wood from a sustainably harvested source.
- Where the sill plate is at least 18" above grade, a metal termite shield may be used in lieu of a treated sill plate. Verify acceptability with local code officials.

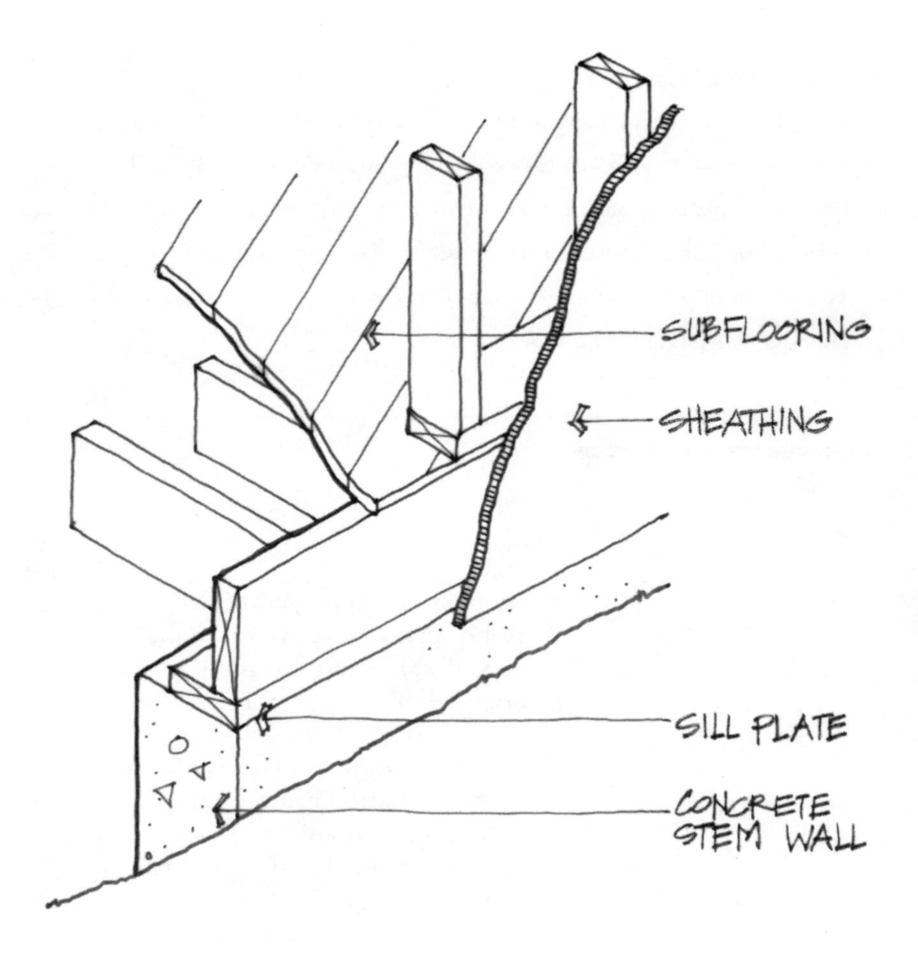

Wood frame construction showing sill plates.

Framing

Wall Framing
Where wall framing is used, follow the guidelines for wood selection and storage in this Division.

Roof and Floor Framing
Solid beams, round logs, or 2x joisting are commonly used for shorter roof spans. Manufactured trusses have several advantages over solid lumber when the framing is not exposed and the spans are large. They are less expensive, use wood more sparingly, have greater span capabilities, provide a deep pocket for roof insulation, and can be fabricated with a built-in slope for flat roof application.

Truss joists, commonly called TJIs, are manufactured beams containing either plywood or dimensional lumber for top and bottom chords, and either plywood or pressboard for the webs. Because they are a very cost-effective way to frame large spans, truss joists are therefore widely used in residential construction. Although the members have been subjected to high heat during manufacturing—which reduces the quantity of volatile organic compounds (VOCs) in their webs—some formaldehyde still remains. In new home construction, the cumulative effect of several low emissions can add up to unacceptable levels. For this reason, we recommend the following treatment.

Truss Joints

- Truss joists (TJI) webs shall be thoroughly sealed prior to installation with **BIN Primer Sealer**. (Refer to Chart 9.1, "Resource List.")

Another option is to use open web roof trusses with dimensional lumber for the top and bottom chords and webs, thereby avoiding the use of pressboard entirely.

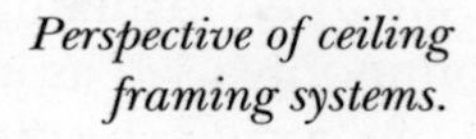
Perspective of ceiling framing systems.

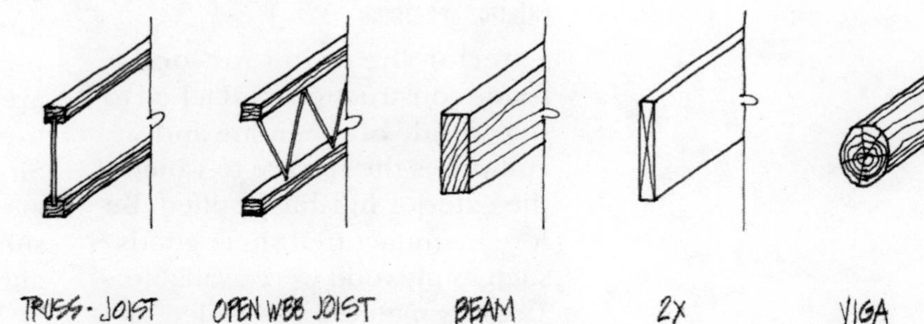

Sheathing

Subflooring

Interior grade plywood and particle board are typically used for subflooring in standard construction. Urea formaldehyde glues are used to bond the wood during manufacturing. This is a concentrated, volatile form of formaldehyde which contributes significantly to indoor air pollution. The subflooring is then attached to the framing underneath with solvent based glues that will also contribute to the pollution level.

Solid wood solutions, as well as the cementitious subfloor sheeting more commonly used in commercial building, can be considerably more expensive. Exterior grade plywood can substitute for interior grade plywood for only a small increase in cost. While exterior grade plywood contains less volatile phenol based formaldehyde glues, it will still release significant amounts of formaldehyde into the air when new. Airing out the wood by cross stacking it on site is better than installing it immediately after delivery. Sealing the wood after it has been aired out will provide the most protection against toxic fumes.

Subflooring

The use of subflooring materials such as interior grade plywood, pressboard, or OSB containing urea formaldehyde glues is prohibited.

The following subflooring materials are acceptable:

- 1x finish floor boards laid parallel to walls over 1x subfloor laid diagonally to walls. (Note: This may be a good solution when a finished wood floor is desired.) Verify for proper span conditions with the architect or engineer.
- 2x4 or 2x6 solid wood tongue and groove
- Exterior grade plywood that has been aired out and then sealed with **BIN Primer Sealer** or another acceptable sealer on all six sides.
- Subfloor adhesive must be solvent free. (Refer to Chart 6-2, "Resource List" for suitable subfloor adhesives.)

Exterior Sheathing

Exterior sheathing in wood frame construction is attached to the outside of the frame and comprises the surface to which the exterior finish is applied. Before manufactured sheet goods such as plywood were available, 1x or 2x material was nailed to the studs for this purpose. In standard wood frame construction today, exterior grade plywood or OSB (oriented standard board, also known as waferboard) is typically used as exterior sheathing at corners where sheer strength is required. These materials contain varying degrees of formaldehyde and isocyanates and do not have the longevity of solid wood products.

Many problems regarding the use of OSB in roof and wall sheathing have recently been identified. In fact, one prominent manufacturer has recently been subjected to a class action suit. When the board gets wet it is vulnerable to fungus invasion and rapidly deteriorates. Asphalt impregnated fiberboard or asphalt sheathed insulating board are commonly used as infill between the corner shear panels. Since asphalt is a known carcinogen, we believe that any exposure level is too high when other alternatives exist.

When vapor barriers are used in the wall composition, sheathing material will not have as great an impact on the indoor air quality as will the materials exposed to the interior. Moreover, the sheathing will have had several weeks in place to air out before it is covered up. In a permeable or "breathing" wall system, where vapor barriers are eliminated with the intent of allowing slow air exchange through the wall, the type of exterior sheathing must be more carefully considered both in terms of permeability and harmful chemical content. The following may be included in your specifications to reduce the pollution generated by exterior sheathing.

Exterior Sheathing

The following products are unacceptable for exterior sheathing:

- products containing asphalt
- odorous foam insulation boards
- pressure treated plywood

The following products and methods are acceptable for exterior sheathing:

- 1x recycled lumber laid diagonally with diagonal metal or wood bracing as structurally required. (A more labor-intensive and expensive solution, this option is most suitable for "breathing wall" frame applications.)
- CDX plywood. Purchase plywood as far in advance of installation as possible and stack it to allow air flow on all sides of each sheet while protecting it from moisture damage.
- **Fiberbond**: A fiber reinforced gypsum sheathing that does not contain asphalt which can be used for exterior sheathing.

Roof Sheathing

Roof sheathing is placed on top of roof framing members and under the roofing. As with exterior sheathing, exterior grade plywood or OSB is most commonly used for this purpose. Unlike wall sheathing, roof sheathing will be exposed to higher temperatures and will therefore be subject to more intense offgassing. Roof sheathing usually has less time to air out in situ since it is roofed over as soon as possible to avoid water damage from precipitation. We therefore recommend that plywood, if used, be treated as specified below. We do not recommend OSB because it can develop mold and deteriorate more rapidly if it happens to get wet.

When roofing members are exposed to the interior, as is often the case when beams or vigas are used, then solid wood planks or tongue and groove sheathing is commonly used.

For sloped roofing, when structural conditions permit, purlins or skip sheathing may be acceptable. Purlins are wooden members spaced to receive metal roof panels, while skip sheathing consists of solid wooden members spaced more closely together for shingle and tile roof applications. Both purlins and skip sheathing eliminate the need for sheeting and provide a good opportunity for ample air movement to ventilate the roof space.

Where a continuous vapor barrier is installed between the framing members and living space, the choice of sheathing material is less crucial.

Consider the following guidelines for inclusion in your specifications.

Roof Sheathing

- The use of solid wood boards, tongue and groove board, solid wood skip sheathing or purlins is preferred.
- CDX plywood should be purchased as far in advance as possible to allow time for aeration. Provide protection against moisture damage.
- Provide a continuous vapor barrier between the plywood sheathing as outlined in "Division 7 - Thermal and Moisture Control," or seal the plywood on all edges and surfaces with an acceptable vapor barrier sealant after it has been aerated. (Refer to Chart 9-1, "Resource List").

Finish Carpentry

Many manufactured composite board products designed for interior use contain urea formaldehyde binders. They offgas formaldehyde for many months, contributing significantly to the indoor pollution level. In standard construction these interior grade composites are used in many applications including cabinetry, shelving, and trim. They should not be used in a healthy home.

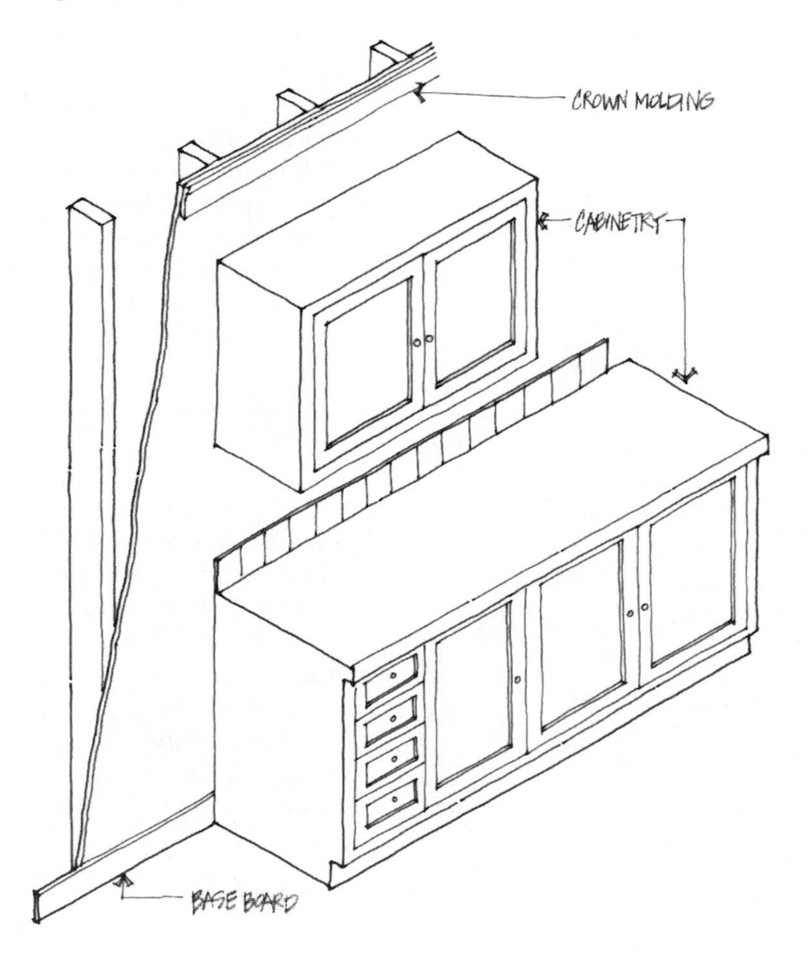

Perspective of finish carpentry, including cabinets, shelving, and base trim.

Finish Carpentry

- No sheetgood containing urea formaldehyde shall be used.
- Exposed interior finish wood shall be comprised of solid wood and finished with an acceptably low VOC finish as specified in "Division 9 - Finishes."
- Where sheetgoods are used choose one of the low emission boards listed below under cabinetry carcass options, or exterior grade plywood that has been aired out, then thoroughly sealed on all edges and surfaces with an acceptable vapor barrier sealant and finished with one of the paints specified in "Division 9 - Finishes."

Cabinets

Although the drawers and doors on cabinetry are often made of solid wood, the boxes are usually composed of particle board, interior grade plywood, or melamine which has a particleboard core that is exposed where holes have been drilled for adjustable shelving. Cabinets are most often finished with solvent based finishes that may also outgas high levels of VOCs for several months.

Because standard cabinetry contributes significantly to poor indoor air quality, it is not acceptable in the healthy house. You will pay more for healthier cabinets, but in terms of indoor air quality, this is money well spent. If your budget is tight, we would suggest that you design strategies that will reduce the amount of cabinetry necessary. For example, you may choose to consolidate a portion of kitchen storage in a pantry area, or you may choose to use attractive solid wood open shelving for dishes or cookware to substitute for cabinetry.

As cabinetmakers are becoming more familiar with the need for healthier cabinetry, and as low VOC finishes and materials are becoming available, the gap in pricing between standard and healthy cabinets is decreasing. Healthier finishes for cabinetry will be discussed in "Division 9 - Finishes." Appearing below is a list of healthier solutions for box construction from which to choose.

Cabinetry Carcass Options

- **Architectural Forest Enterprises**: Veneered hardwood products
- **Cervitor**: Distributors of metal cabinetry with a baked-on enamel finish that may be used with metal or solid wood doors and drawers
- **Medex or Medite II**: A medium density fiberboard manufactured without formaldehyde
- **Multi-core**: A low emissions plywood which comes with a variety of hardwood veneers
- **Neff Cabinets**: Manufactured cabinets with a 98% reduction in formaldehyde content
- **Panolam**: Manufacturers of a melamine sheeting with a Medex core
- Plate glass shelving
- For all options, specify the use of a solvent free carpenter's glue.

Countertops

The ideal countertop material for a healthy home would have a solid, nonporous surface that is stain and scratch proof. It would be attachable by mechanical means directly to the cabinet carcasses, thus avoiding the need for underlayment and adhesives. It would be beautiful, inexpensive, and manufactured in a variety of colors. Unfortunately, all of these characteristics are not found in combination in a single countertop option. Chart 6-1, "Countertop Comparisons," reviews the most common countertop materials and outlines specification concerns for each.

Case Study:
The radioactive countertop

John Banta was called to the home of a woman who was employed as a cook by the television industry. Her task was to prepare samples of the same recipe in various stages of preparation, from raw ingredients, to oven-ready mixture, to finished product. The prepared foods were then delivered to the television studio so the recipe could be demonstrated by a celebrity on a culinary arts program.

During the investigation, John discovered that his client was being exposed to an unexpected occupational hazard. The orange colored tile used for her counter was glazed with uranium oxide, a highly radioactive substance which was making the numbers on the Geiger counter spin too fast to count. For over 30 years this woman had worked at a radioactive counter slicing, dicing, mixing, and arranging her creations.

When she learned of the radioactivity, the client revealed to John that she had recently had a pre-cancerous lesion removed from her intestines. Her surgical scar was located at the level of the counter where it pressed against her while she cooked. The client was advised to have her counter top replaced. Her physician concurred.

Chart 6-1: Countertop Comparisons

Type	Relative cost	Advantages	Disadvantages	Comments	Specify
High pressure laminates (e.g., Formica, Wilsonart)	Lowest initial investment	• Wide variety of colors, patterns, textures, and sheens • Low cost • Seamless surface	• Glued to particle board with toxic glues • Particle board outgasses formaldehyde • Not stain or acid resistant • Will scratch • Cannot be resurfaced • Short life; deteriorates quickly if the particle board gets wet	• Not a good choice for a healthy home	• Fasten to carcass with mechanical fasteners • Seal all exposed edges and surfaces of particle board with foil or one of the vapor barrier sealants listed in Division 9 – Finishes
Solid surfaces materials (e.g., Corian, Avonite, Swan-stone)	Expensive	• Nonporous • Sanitary; integral bowls and rolled back splashes are easy to clean • No substrate needed for most • Scratches and stains can easily be sanded • Attractive marble and granite like surfaces • Can be mechanically fastened	• Color selections are limited • Can be more expensive than granite or marble	• Select a type that does not require substrate	• Fasten to carcass with mechanical fasteners
Tile	Can be moderate	• Hard, scratch resistant surfaces • Large variety of sizes, colors, textures to choose from	• Grout joints are subject to staining and mold and bacteria growth • Glazes may contain heavy metals or be radioactive • Tiles can crack or chip under heavy impact	• Choose the largest tiles available to reduce the number of grout joints • Choose presealed tiles • Choose commercially rated tiles • Choose tiles requiring the narrowest grout joints • Choosing porcelains with integral color will disguise chips	• Follow recommendations in the tile section of "Division 9 – Finishes" for underlayment, tile setting and grout sealing
Butcher block	Moderate	• Warm, inviting aesthetics • Natural material • Can be refinished by sanding • Does not require underlayment • Can be mechanically fastened	• Porous surface stains easily and can harbor mold and bacteria growth	• Four-joint seams may be glued with formaldehyde based adhesives	• Finish with odorless, nontoxic oil such as walnut oil • Fasten to carcass with mechanical fasteners • Shall be formaldehyde free

Type	Relative cost	Advantages	Disadvantages	Comments	Specify
Solid sheet granite	Expensive	• Wide selection of very beautiful stones • Hard, scratchproof, stain resistant surface that will last forever • Solid seamless surface • Can be mechanically fastened or glued with silicone • May not require substrate	• May be cost prohibitive • Surface must be finished with impregnating finish	• Oil or butter left on surface will stain it • Must check for radioactivity	• Submit MSDS for surface impregnator • Shall be mechanically fastened or fastened with 100% pure silicone caulk (aquarium/ food grade)
Granite tile	Moderate	• Can resemble granite but is less expensive than slab • Very thin grout joints can be sealed with transparent silicone • Mar and scratch resistant	• Requires epoxy type glues to set • Requires underlayment	• Oil or butter left on surface will stain it • Must check for radioactivity	• Submit MSDS for surface impregnator • Shall be mechanically fastened or fastened with 100% pure silicone caulk (aquarium/ food grade)
Stainless steel	Expensive	• Nonporous, nonstaining • Easily cleaned continuous surface	• Aesthetically cold • Most suitable for contemporary kitchens • Thinner gauges require underlayment • Noisy • Must be special ordered	• Conducts electricity • Proper ground fault interruptors are essential to prevent potential electrocution	• Use formaldehyde free underlayment and mechanical fastening

Chart 6-2: Resource List

Product	Description	Manufacturer/Distributor
9400 W Impregnant	Solvent free, water repellant coating that allows wood to breathe while providing ultraviolet, mildew, and frost protection	Palmer Industries, Inc.; 10611 Old Annapolis Rd.; Frederick, MD 21701; (888)685-7244, (301)898-7848
ACQ Preserve	Pressure treatment for wood that uses alkaline copper quat containing no known carcinogens or EPA listed hazardous compounds. Not widely distributed.	Chemical Specialties Inc.; One Woodlawn Green, Suite 250; Charlotte, NC 28217; (800)421-8661
AFM Safecoat Almighty Adhesive	A low odor, nontoxic, water based adhesive for gluing wood and wood laminates.	AFM (American Formulating and Manufacturing); 350 West Ash Street, Suite 700; San Diego, CA 92101-3404; (800)239-0321, (619)239-0321
Architectural Forest Enterprises	Hardwood plywood veneers from certified sources over a core of Medite II.	1030 Quesada Av.; San Francisco, CA 94124; (800)483-6337, (415)822-7300
Bio Shield Hardwood Penetrating Sealer	As above but a more diluted solution for use on hardwoods.	Eco Design/Natural Choice; 1365 Rufina Circle; Santa Fe, NM 87505; (800)621-2591, (505)438-3448

Product	Description	Manufacturer/Distributor
Bio Shield Penetrating Oil Sealer #5	Undercoat for soft wood, best used with a finishing stain for UV protection. Free of petroleum distillates and mineral spirits.	Same
Bio Shield Transparent Wood Glaze	Interior/exterior wood finish with ultraviolet protection.	Same
Bora-Care	Designed to penetrate and protect all types of wood from wood boring insects. Contains disodium octaborate tetrahydrate in ethylene glycol carrier. Water based solution requiring paint or sealer on top.	Nisus Corporation; Knoxville, TN Distributed by Nontoxic Environments Inc.; P.O. Box 384; Newmarket, NH 03857; (800)789-4248, (603)659-5919
Old Growth Aging and Staining Solutions for Wood	Wood is treated with a nontoxic mineral compound and then with a nontoxic catalyst that binds the natural mineral colors to cellulose creating an aged patina. It imparts antimicrobial and antifungal properties to the wood while the pigments provide UV protection.	CrossLink, LLC; P.O. Box 1371; Santa Fe, NM 87504-1371; (505)983-6877
Cervitor	Metal kitchen cabinetry.	Cervitor Kitchens Inc.; 19775 Lower Azusa Road; El Monte, CA 91731-1351; (800)523-2666, (818)443-0184
Eco Timber International	Source for sustainably harvested wood.	1020 Heinz Avenue; Berkeley, CA 94710; (510)549-3000
Elmer's Carpenter's Glue	Solvent free glue.	Borden Inc.; 180 Broad Street; Columbus, OH 43215; (800)426-7336, (800)848-9400 Available in many retail outlets.
Fiberbond	A reinforced gypsum sheathing used as a backer for exterior finishing.	Louisiana Pacific; 111 SW Fifth Ave.; Portland, OR 97204; (800)579-8401
Forest Trust Lumber Brokerage	Lumber broker for small independent producers in New Mexico; lumber sold to contractors and homeowners. Forest Trust has developed its own certification program and is currently working with Rainforest Alliance to obtain Smart Wood certification.	Todd Myers; P.O. Box 519; Santa Fe, NM 87505; (505)983-3111
Guardian	Borate based wood preservative.	See Shellguard.
Livos Donnos Wood Pitch Impregnation	A penetrating preservative for exterior woodwork in contact with moisture. Made of natural ingredients using plant chemistry.	Eco Design/Natural Choice; 1365 Rufina Circle; Santa Fe, NM 87505; (800)621-2591, (505)438-3448
Livos Dubno Primer Oil	A penetrating oil primer for use as an undercoat on exterior wood.	Same
Medex	Formaldehyde free, exterior grade, medium density fiberboard.	Medite Corporation; P.O. Box 4040; Medford, OR 97501; (800)676-3339, (503)773-2522
Medite II	Formaldehyde free, interior grade, medium density fiberboard.	Medite Corporation; P.O. Box 4040; Medford, OR 97501; (800)676-3339, (503)773-2522

Product	Description	Manufacturer/Distributor
Multi-core	Hardwood veneered plywood panels with low formaldehyde emissions. Suitable for cabinetry.	Weldwood of Canada, Inc.; 2000 Argentina Road; Mississauga, Ontario, Canada L5N 1P7; (905)542-2700, (888)566-4522
Neff Cabinets	High quality manufactured cabinets with low formaldehyde emissions. Boxes are made of phenolic glued plywoods. Solid wood doors can be ordered unfinished.	Neff Kitchen Manufacturers; 6 Melanie Drive; Brampton, Ontario, Canada L6T 4K9; (800)268-4527, (905)791-7770
OS Wood Protector	A penetrating, natural oil based wood preservative with zinc oxide. For use on wood exposed to high humidity and moisture to prevent mold and mildew. Does not prevent insect infestation.	Ostermann & Scheiwe; P.O. Box 669; Spanaway, WA 98387; (800)344-9663
Panolam	A melamine board thermally fused to a medex core.	Panolam; 3030 Calapooia; Albany, OR 97321; (888)726-6526, (541)928-1942
Plaza Hardwood, Inc.	Source for sustainably harvested and recycled wood.	Toni and Paul Fuge; 5 Enebro Court; Santa Fe, NM 87505; (800)662-6306, (505)466-7885
Polyshield	A tough, super hard, nonyellowing polyurethane that is UV stabilized and UV stable. Use with Hydrocote stains. Can be used with Hydrocote Ultraviolet Light Absorber Blocker to increase UV resistance.	The Hydrocote Company Inc.; P.O. Box 160; Tennent, NJ 07763; (800)229-4937, (908)257-4344
Quality Wood Products	Sustainably harvested wood.	Harry Morrison; P.O. Box 98; Chama, NM 87520; (505)756-2744
Rainforest Alliance	Nonprofit organization that has set the standard for the "Smart Wood" program. Provides sources of sustainably harvested wood. Ask for Smart List.	65 Bleeker Street; New York, NY 10012-2420; (888)MY-EARTH, (212)677-1900
Shellguard	Borate based wood preservative.	Perma-Chink Systems, Inc. 1605 Prosser Road; Knoxville, TN 37914; (800)548-3554
Solvent Free Titebond Construction Adhesive	A multi-purpose adhesive for a variety of porous surfaces, including plywood and wood paneling.	Franklin International; 2020 Bruck Street; Columbus, OH 43207; (800)347-4583
Titebond Solvent Free Subfloor Adhesive	A multi-purpose adhesive for a variety of porous surfaces.	Franklin International; 2020 Bruck Street; Columbus, OH 43207; (800)347-4583
Timberline 2051	Wood flooring adhesive.	W.F. Taylor Company, Inc.; 11545 Pacific Avenue; Fontana, CA 92337; (800)397-4583, (909)360-6677
Timbor	Disodium, octaborate wood preservative protects against termites, fungus, and wood boring beetles.	U.S. Borax Corporation; 26877 Tourney; Valencia, CA 91355; (800)553-4872, (310)522-5300
WEATHER PRO	A water based semi-transparent wood stain and water repellant.	Okon Inc.; 6000 West 13th Ave.; Lakewood, CO 80214; (800)237-0565, (303)232-3571

Division 7 - Thermal and Moisture Control

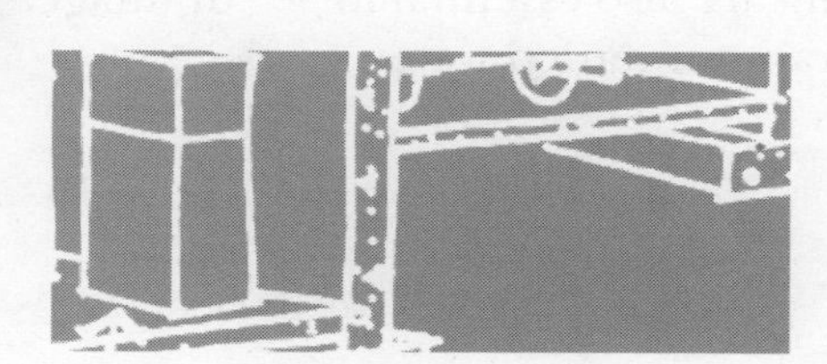

Dampproofing and Waterproofing

Dampproofing is used to form a water resistant barrier between the home's stem walls and the earth wherever the living space of the home is below grade. Along with proper grading, dampproofing is protection against the migration of moisture through the wall. This migration can cause damp plaster or gypboard, and the subsequent invitation for mold growth. The use of asphaltic and bituminous tar mixtures for this purpose is standard practice. These materials are known carcinogens and should not be used in a healthy house. There are several other readily available products made for this purpose which are more healthful choices. The following products may be specified for dampproofing foundation walls or other walls adjacent to soil.

Dampproofing and Waterproofing

Cementitious Dampproofing

- **Thoroseal Foundation Coating**: A cementitious waterproofing for concrete surfaces

Fluid Applied Dampproofing

- **AFM Safecoat DynoSeal**: A flexible, vaporproof barrier
- **AFM Safecoat DynoFlex**: A topcoat for use over DynoSeal

Bentonite Waterproofing

- **Volclay**: A self-healing bentonite based waterproof panel

We have seen many mold problems develop in below grade areas of a home. Dampproofing is only one component in the creation of a tight water barrier. Proper drainage backfilling and final grading are also essential in order to drain unwanted water away from the wall. Conscientious and thorough workmanship are of the utmost importance. Appearing below are sample specifications for the proper installation of perimeter drainage.

Installation of Perimeter Drainage

A drain system shall be installed around the perimter of the foundation footing. The drainage system shall consist of the following items:

1. Positive drainage away from the building along the entire perimeter.
2. Dampproofing of all exterior wall surfaces that are below grade or in contact with soil. (Refer to "Dampproofing and Waterproofing" sample language above for acceptable product list.)
3. Free draining system or free draining back fill that allows for water to drain down from the underground wall to the french drain system.
4. French drain or perforated pipe drainage system at the base of the footing to carry water away from the underground portion of the structure.

- Dampproofing shall be carefully applied according to the manufacturer's directions to cover all below grade surfaces to form a watertight barrier.
- Care shall be taken during back filling and other construction to prevent damage to the dampproofed surface.
- A free draining backfill of 3/4" minimum crushed stone or gravel that is free of smaller particles shall be used to line and fill the excavation for all below grade walls. (An engineered drainage system may be substituted. These systems frequently incorporate perimeter insulation with the drainage. These systems must be installed in strict compliance with manufacturers' specifications in order to properly drain.)
- If crushed gravel or stone is used for the free draining back fill, the top should be covered with a low permeability material such as clay, soil, or concrete, and sloped away from the foundation in order to minimize the amount of water that will be handled by the french drain system.
- A french drain shall be installed so that all perforated pipes are located below the level of the bottom surface of the basement floor slab.
- French drain perforated pipes shall be installed with the holes down to allow water to rise into the pipe. If holes are present in more than one side of the pipe, at least one set of holes shall face downward.
- The perforated pipe shall be surrounded and set in a minimum 2" depth bed with a minimum 3/4" of crushed stone free of smaller particles to prevent clogging.
- The perforated pipe and crushed stone shall be surrounded by a filter membrane to prevent adjacent soil from washing into and clogging the french drain system.
- French drains shall be sloped downward a minimum 1/4" per foot of run and connected to daylight. If the french drain cannot be connected to daylight, it may have to be connected to an underground engineered collection pool, a sump pump, or a storm sewer system. This situation is not ideal because sump pumps can fail and storm sewers can back up. If these problems are not quickly corrected, water damage may result. If the storm sewer is connected to the sanitary sewer, any backup may also result in sewage on the exterior side of underground walls.

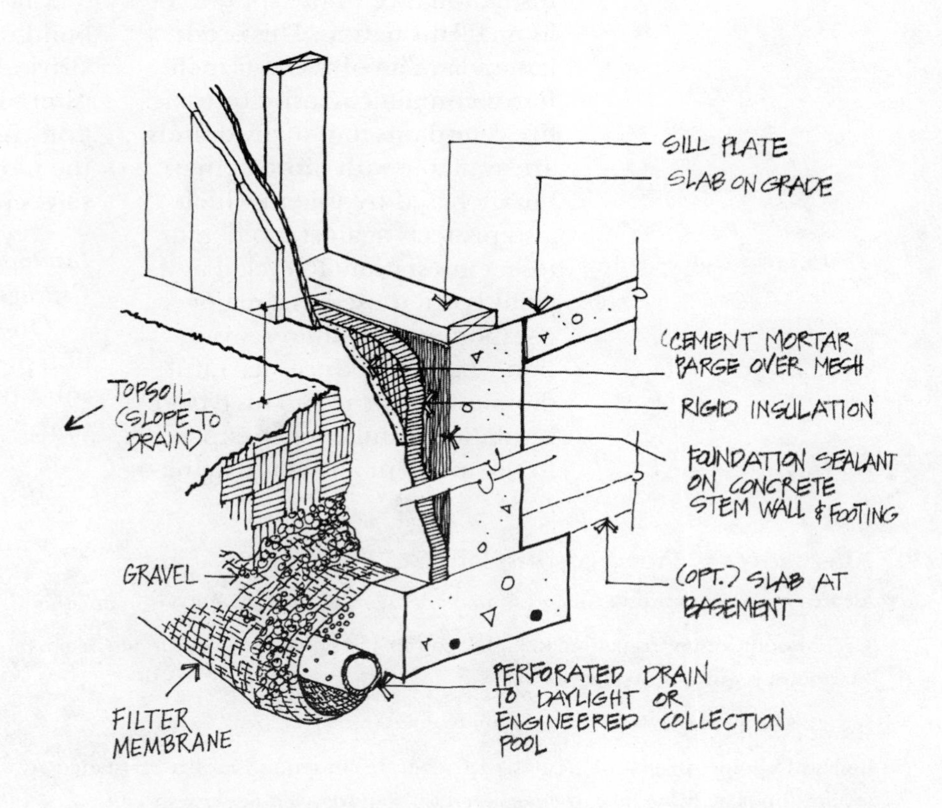

Proper waterproofing.

Thermal Protection

Building Insulation

Ninety percent of the homes in the United States are insulated with fiberglass insulation. There has been much debate as to whether or not fiberglass is a human carcinogen. Whatever the case may be, fiberglass is by no means a healthful substance. Fiberglass insulation releases both particulate matter and gaseous contaminants including formaldehyde and asphalt (if asphalt backed). There are numerous reports linking fiberglass to pulmonary disease in production workers and installers. Although more healthful alternatives exist, they are generally more expensive and may not be as readily available. The cost of insulation, however, comprises a very small proportion of the overall building cost. Even doubling this figure will not constitute a large increase in cost per square foot for the overall construction of your home.

One of the more reasonably priced alternatives to fiberglass insulation is cellulose spray-in or loose fill insulation. This product has an R-value of ±35 per inch. It can contain corrosive or toxic fire retardants, but many brands are available with more benign borate based treatment which also protects against mold and insect infestation. Recycled newsprint is often used as a major component of cellulose insulation which introduces harmful dioxins into the mix. This type of insulation should not be exposed to the ambient air. Some manufacturers provide virgin or cardboard content instead.

Choosing one of the alternate building systems discussed in "Division 4 - Masonry and Other Alternatives to Frame Construction" is another option, where the more massive walls themselves provide the insulation.

Insulation in Cavity Walls, Joisted Ceilings and Joisted Crawl Spaces

One of the following insulation products may be specified to substitute for fiberglass insulation.

Alternatives to Fiberglass Insulation

- **Air Krete***: A cementitious magnesium oxide foam insulation which is foamed in place.
- **Greenwood Cotton Insulation**: Recycled cotton and polyester batts applied in a manner similar to fiberglass batts, but do not require protective gear.
- **Icynene Insulation System***: Insulation foamed in place.
- **Celbar**: Cellulose insulation treated with a borate compound for fire resistance. Available in loose fill or spray-in application. The loose fill can be ordered without recycled newspaper content.

* Wet applied insulation shall be thoroughly dry prior to application of air barrier (assuming one is used).

Finally, if fiberglass batt is your only option for insulation, then specify one of the less toxic applications listed below.

If Using Fiberglass Batt

1. Choose undyed yellow or white batts without backing as available through **CertainTeed** or an equivalent. Install an acceptable air barrier, as specified below, on the inside face of the studs, joists, or rafters between the insulation and the living space.
2. Purchase fiberglass batts that come prewrapped in polyethelene bags. Most fiberglass insulation companies are now providing this wrapping. The bags have limited application, however, because they must be cut open and trimmed wherever spacing is irregular.
3. Specify **Miraflex**, an undyed product made without binders and encased in a polywrap. Application is limited because the product is currently available only in 16" and 24" wide R-24 packages.

Insulation Over Beams and Vigas

Where structural members of the ceiling are exposed, as is common with Santa Fe style homes, the air space between the structural members is not available to receive insulation. This poses a problem for roof insulation. Various available tapered insulation systems are designed to go over the exposed ceiling decking. The less toxic of such alternatives tend to be expensive. A more cost effective solution is to build in a cavity area over the existing exposed ceiling and insulate with one of the abovementioned products.

Insulation Around Windows and Doors

Insulation around windows and doors is an important factor in energy conservation. In standard construction an expandable foam spray is applied which typically contains hydrochlorfluorocarbons (HCFCs). Although such spray is excellent in terms of sealing, noxious fumes are introduced into the building envelope, as well as damaging the Earth's ozone layer. If standard foam is used, we recommend sealing it with an air barrier tape as specified below. A healthier alternative is to use Cord Caulk, an acrylic yarn saturated with an adhesive wax polymer.

Insulation Around Windows and Doors

- Use polyurethane foam with no CFCs, HCFCs, or formaldehyde. Cover with **Polyken Tape +337** where exposed to the interior.
- Apply **Cord Caulk** in all locations where foam caulking would normally be applied. Press by hand into the area between the windows and doors and the framing. The product may also be used under sills.

Air Barrier

We recommend using an air barrier in situations where separating the interior space of the home from toxic substances is necessary. For example, if fiberglass batt insulation is the only available or affordable choice for insulation, a properly installed air barrier between the insulation and the living space will help reduce the infiltration of toxic substances. It is almost impossible to avoid punctures, given the complexities of construction and the number of materials that must be mechanically fastened together. Appearing below are instructions that can be included in specifications if you install air barriers.

Air Barrier on Frame Walls, Ceilings and Frame Floors

If fiberglass batts or other offgassing materials are used in wall construction, the air barrier shall be applied on the inside face of studs, joists, or rafters (warm side) just prior to the application of the interior facing. After applying the acceptable air barrier (see list below), seal with 100% silicone caulk and foil tape. Staple the barrier in pieces that are as large as possible over the insulation and attach them to the window and door jambs with staples and approved caulk to form a complete seal. Caulk at all wall openings such as plumbing and electrical boxes. Tape or caulk all seams and joints. Caulk all electrical boxes at the hole where the wire comes through, or else purchase gasketed boxes.

NOTE: This method is not intended for use in hot, humid climates, especially where air conditioning is in use. Condensation on the insulation side of the barrier may occur which causes hidden water damage and microbial growth. When an air barrier is to be applied, use only unbacked insulation.

Acceptable Air Barriers

The following products generate little or no emissions and are suitable for air barriers.

- **Cross Tuff**: Cross laminated, polyethylene sheeting. If you specify "for a healthy house," the manufacturer will then incorporate two additional processes.
- **Denny Foil Vapor Barrier***: Virgin kraft paper laminated with foil containing sodium silicate adhesive on both sides.
- **Insulfoil***: Foil laminated on two sides of kraft paper with nontoxic adhesive.
- **Tu-Tuf 3** or **Tu-Tuf 4**: High density, cross laminated, polyethylene sheeting.

*Not suitable for areas that may get wet.

Roofing

A well-sloped roof with a sizable overhang is preferable over a flat or low sloped roof for a healthy home. Listed below are explanations for this preference.

- The roof overhang plays an important role in protecting the walls and foundations from water damage by directing water away from the building.
- Inert roofing materials are readily available and are standard products for sloped roof construction, whereas they are an exception in flat or low-sloped residential roof construction.
- Sloped roofs shed water quickly, whereas water will puddle and linger on poorly constructed flat roofing.
- Flat roofs have a higher failure rate which may lead to devastating mold problems.
- Overhangs can be sized to suit the solar conditions in your region to provide shade in the summer while allowing maximum heat entry in the winter.

Straw bale building with metal roof. (architect: Paula Baker. Contractor: Living Structures, Inc.)

Sloped Roofing Materials

Asphalt based rolled roofing and shingles will offgas when heated by the sun and should be avoided. Clay tile, concrete tile, metal, and slate are all healthy, longlasting slope roof solutions. Wood shingles can be a good roofing material where fire danger is low and if rot resistant woods such as cedar are used. Zinc or copper strip used at the ridge will wash wood shingles with preservatives every time it rains. Availability of these roofing materials varies from region to region.

In many cases, roofers will want to install an asphalt based felt paper under healthier roofs. If you are unable to locate an experienced roofer who will warranty the work without this liner, then it is important to seal the roofing material from the structure by using a suitable air barrier.

Membrane Roofing

Membranes for flat roofing are more problematic. These roofs are in fact more accurately described as having a very low slope, usually 1/4" per foot or less. Tile shingles, and most metal applications which depend on rapid water runoff, will not hold up under standing water conditions and are not suitable for low slope applications.

Tar and gravel roofing is the most common and least expensive material available for flat roof applications, but we do not recommend it. A tar and gravel roof will emit volatile organic compounds (VOCs) from asphalt, benzene, polynuclear aromatics, toluene, and xylene. It will continually outgas when heated by the sun. Some of these vapors will inevitably find their way into the living space and degrade air quality. Eventually the roof will outgas to the point where it does not adversely affect indoor air quality. The average tar and gravel roof is guaranteed for only two to five years and will require replacement in less than 10 years. Since most people are not in a position to move out for several weeks when their roof requires replacement, they will be exposed to high levels of toxic fumes every time the roof is repaired or replaced. Chemically sensitive individuals often have difficulty tolerating a tar and gravel roof that is less than one to two years old.

Toxicity is not the only health concern that should be considered when choosing a product. Many unhealthy and persistent mold and mildew infestations begin with an undetected roof leak. No type of roofing installation is foolproof, but the use of high quality roofing materials and skilled installers will reduce the risk of leakage.

Although other solutions are typically more expensive than tar and gravel, you must carefully weigh both life cycle and health costs when making a roofing choice. Single ply membranes such as **Brae Roof** contain asphalt and will also outgas to a certain extent when heat is applied to fuse the membrane during application, but they are fairly stable once installed. They also carry a longer warranty period. Certain single ply membranes can be repaired by welding patches onto the existing roof, thereby extending the roof life for many years. There are also several roll-on paint applications that do not require roofing contractors for application or repair.

Several new roofing products have recently been introduced. It is especially important with roofing materials that the manufacturer's instructions for installation and warranty criteria be carefully followed. The following products have been selected from among the many available alternatives to tar and gravel roofing for flat roof applications.

Sources for Flat Roofing

- **Resource Conservation Technologies, Inc.**: An acrylic polymer paint- or roll-on system that uses titanium dioxide with propylene glycol and contains no toxic dispersants or tints.
- **Brae Roof**: Specify torch-down application; otherwise the roof may be attached using hot tar which will outgas indefinitely when reheated by the sun.
- **Hi Tuff EP**: A heat weldable rubber roofing membrane.
- **Mirrorseal**: A single ply, fluid applied roofing system.
- **Thermomaterials**: Paint-on roofing.

Joint Sealants

Many solvent based caulking compounds are formulated with hazardous solvents such as acetone, methyl ethyl acetone, toluene, and xylene. They are moderately toxic to handle and may offgas for extended periods of time. Appearing below are suggested options.

Joint Sealants for Exterior Use

On the exterior of the house, use 100% silicone caulk, aquarium grade, of any brand. (Be sure to read labels because some are labeled "pure silicone," but contain other ingredients.) Commercially available safer caulks include the following:

- **AFM Safecoat Caulking Compound**: Water based elastic emulsion.
- **GE 012**: Clear Silicone Sealant, or **GE 5091; Silicone Paintable Sealant**.
- **Dow Corning**: 100% Silicone Sealant or Silicone Plus.
- **Lithoseal Building Caulk**: Urethane modified polymer that is inert once cured.
- **Phenoseal "Surpass" Caulk** and **Sealant**, **Vinyl Adhesive Caulk**, **Valve Seal**: Line of water based sealants and caulks.
- **Cord Caulk**: This is an organic yarn that has been saturated with an adhesive wax polymer.

Apply according to manufacturer's directions around all openings, such as doors, windows, below exterior door thresholds, and wherever necessary to obtain a complete weather tight, exterior building seal.

Radon

Radon is a clear, odorless gaseous by-product of the natural breakdown of uranium in soil, rock, and water. While radon gas dissipates in open spaces, it tends to cling to particulate matter and accumulates when enclosed. Upon inhalation, radioactive particles become lodged in the mucous membranes of the respiratory system. The Surgeon General has stated that radon exposure is second only to tobacco smoke as a cause of lung cancer.

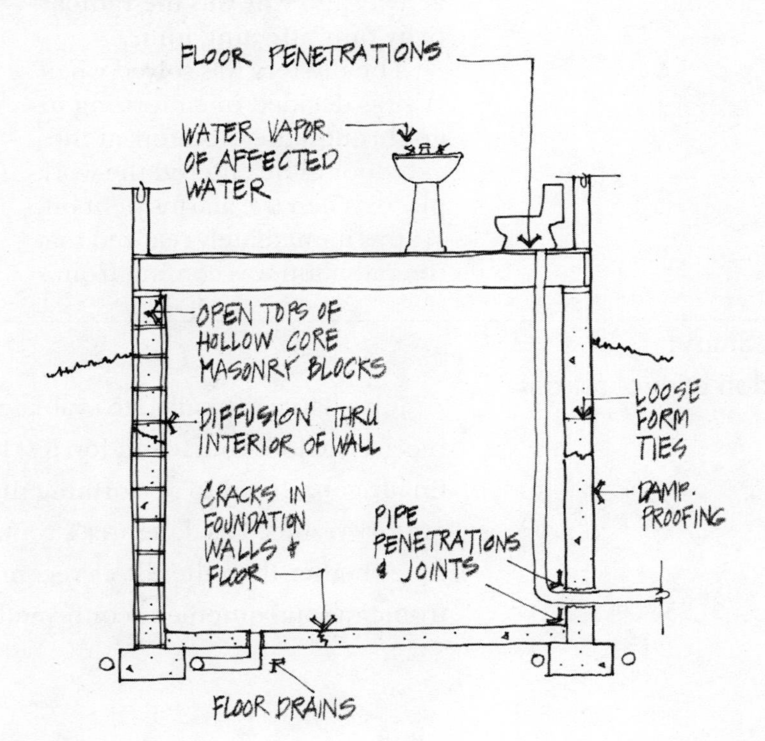

Sources of radon.

How Radon First Came to the Attention of the U.S. Public

Stanley Watras had worked as an engineer for 11 years at a nuclear power plant in Pennsylvania. At the end of each workday, he and other plant employees were checked by a monitor that measured radiation levels. This procedure ensured that they had not been contaminated by unsafe levels of radioactivity while at work.

In December 1984, Stanley suddenly began setting off the buzzers on the radiation monitors as he walked by the machine on his way out of the building. The readings showed high levels of contamination over Watras's entire body. For several days this scenario was repeated with Watras subjected to a lengthy decontamination ordeal. Where was Watras picking up this radioactivity and why was the radioactivity only affecting him?

The mystery was solved when Watras decided one morning to go through the monitors at the exit door as he entered the workplace. When the alarms went off, Watras immediately realized that the radiation was coming from somewhere outside the nuclear power plant. The local electric company sent a team of specialists to Watras's house to investigate. The readings on the Geiger counter showed levels 700 times higher than the maximum considered safe for human exposure.

Researchers concluded that the culprit was radon, a naturally occurring radioactive gas derived from underground uranium. At that time, very little was known about radon and its health effects. The Watras house was used as a laboratory for radon researchers who wanted to learn how radon gets into a house and how to get it out. Low-grade uranium ore was discovered beneath the basement of the structure, in direct contact with the house. The foundation of the house was removed, along with the soil underneath, to a depth of four feet. Ventilation fans were installed to pull radon-laden air out from under the house. Watras and his family were eventually able to move back into their home.

Case Study:
A radon control retrofit

John Banta was called to evaluate a home for radon. The owner had received a do-it-yourself radon test kit as a gift from relatives. When he finally got around to performing the test, he could not believe the laboratory results. His daughter's room registered 24 pico-curies, six times higher than the EPA's recommended action level. John's electronic radon equipment confirmed the test results.

A radon reduction technique called sub-slab suction was proposed to the client which meant that radon would be sucked from under the slab and ventilated to the outside of the home. Holes would be drilled in the downstairs slab so that pipes could be inserted and connected to an exhaust fan, a method frequently used in unfinished basements. Since the owner had just finished installing an expensive marble floor downstairs, he was not willing to accept this proposal.

After some thought, John suggested that the sub-slab suction technique be modified so that the drilling would take place horizontally under the slab through the outside of the hill on which the first floor rested. A company that drills horizontal wells was contracted for the job. The site was surveyed and the drill set to bore just under the foundation. Six evenly spaced holes were bored horizontally all the way under the house. After the drill was withdrawn from each hole, a perforated pipe was inserted which provided a pathway for gas from radon contaminated soil to be sucked from under the home. The owner finished the job by joining all perforated pipes together with solid pipe which he ran into a small shed a short distance from the home. An exhaust fan was connected to the pipe to suck radon to the outside where it dissipated. The pipes were then covered with soil and the area landscaped. The home's radon level was reduced to an acceptable level (around one pico-curie). If the fan is shut off, however, the radon level again begins to climb.

More radon testing was carried out on other buildings located on the property and in the general neighborhood. No other elevated radon levels were found.

Discussion

Radon can exist in isolated spots, depending on underlying geological formations. Some parts of the United States are known to have higher radon levels than others. Homes with basements, cellars, or other subterranean structures are the most susceptible to radon accumulation. Yet even homes with slab foundations and ventilated crawl spaces can have elevated levels. The only way to be certain is through radon testing. In John's experience, radon can almost always be reduced to acceptable levels. When building your home, appropriate techniques can be used to avoid the possibility of radon accumulation.

Radon Mitigation in New Construction

It has been estimated that as many as one in 15 homes in the United States contains elevated radon levels. The EPA recommends remediation at levels higher than 4.0pCi/L (pico-curies per liter of air). Even at 4.0pCi/L there is an increased risk of lung cancer; therefore, reducing radon to levels between 1.0 and 1.5pCi/L is recommended.

Does your building site contain high radon levels? There is no foolproof way to test a building site for radon. The amount of radon coming out of the ground can vary substantially from one part of the building footprint to another just a few feet away. A pocket will affect the entire building envelope once enclosed. Construction methods and pressure dynamics of the finished building will also affect radon levels.

A radon land test kit available through **Airchek** consists of an activated carbon radon collection packet and a tenting device. The detector is exposed as directed in the instructions and then sent to a lab for results. Testing the footprint area in at least two locations is recommended. If the tests indicate a problem, then you know you indeed have a problem. However, if they do not indicate a problem, there is no assurance that the finished home will not have elevated radon levels.

Another source of information is the radon mitigation experts in your area. Someone with experience who has conducted hundreds of tests will be aware of regional trends. For example, in the Santa Fe, New Mexico, area, elevated radon levels are not uncommon in the foothills where one encounters large quantities of solid granite. On the other hand, elevated radon levels have never been found in subdivisions a few miles west of the foothills.

Foundation structure type also affects the amount of radon that will accumulate in a building if radon is present in the soil. Basement construction is the most vulnerable to radon seepage because it has the largest surface area in contact with the soil. Crawl spaces under buildings, especially unvented ones, can store radon gas. The gas is easily transferred to the living space if an effective air barrier separating the living space from the crawl space does not exist. A slab on grade can form an effective barrier against radon, but any cracks, joints, or penetrations in the slab will create routes for radon to enter. Mud floors or other types of permeable floor systems that come into direct contact with the ground are not recommended without the application of supplementary radon mitigation where elevated radon levels are suspected.

Methods of Radon Mitigation

The EPA conducts training programs for contractors. State offices can provide you with the names of contractors who have been trained under the EPA's Radon Contractor Proficiency Program and of those who are RCP qualified. A good strategy for radon mitigation consists of the following three components.

- Blockage of all potential entry routes. Concrete slabs and basement walls must be properly reinforced to prevent cracking. (Refer to "Division 3 - Concrete" for information on concrete reinforcement.) Cold joints, expansion joints, and plumbing penetrations must be sealed with a flexible caulk. (See the "Resources List" below on acceptable caulking.) Special barrier sheeting placed under the slab or under the floor joists in the crawl space will further block radon gas from entering. Basement walls must be thoroughly parged. Slab and block walls can by coated with **AFM Safecoat Dynoseal** to further seal cracks and joints.

- Prevention of negative pressurization of the building envelope. A home that has lower air pressure than the surrounding outside environment will be negatively pressurized. A vacuum is created which will suck air into the building from wherever there happens to be a route of entry. This suction may cause the entry of gases containing radon from the soil which can seep through tiny cracks in the slab, crawl space, or basement walls. It is important to provide sources for the controlled supply of outside air into the home to replace the air lost through various appliances, including furnaces, exhaust fans, and fireplaces. Creating a condition where there is a slight positive pressurization can be an effective means of reducing radon levels. Strategies for providing proper pressurization are discussed in "Division 15 - Mechanical."

- Collection of radon gas from under the building envelope and redirection away from the building. There are several methods for accomplishing this task. Appearing below are two examples that may be included in the specifications, along with instructions for proper installation of barriers and sealants.

Radon Mitigation

Method #1: A 4" layer of aggregate is placed under the building envelope. A 4" diameter perforated pipe is laid in the aggregate through the center of the envelope. The pipe is connected to an unperforated riser tube that vents to the outside. The vent tube acts as a passive radon removal outlet. If radon levels are still unacceptable once the building is completed, a fan can be attached to the vent pipe to actively suction out the gas.

Method #2: In place of aggregate and perforated pipe, **Soil Gas Collector Matting** can be laid on the finished grade prior to pouring concrete. The matting, which is covered in filter fabric, is laid around the inside perimeter of the foundation in a swath about 1 ft wide, and the concrete poured directly on top. The matting is connected to a vertical riser vent which extends through the roof. The natural chimney effect will draw the radon gas upward. If deemed necessary, the system can be adapted for active suction with the addition of a fan once the building is enclosed.

NOTE: In areas with high water tables, consult a geotechnical engineer about proper drainage prior to installation of any radon removal system.

Acceptable Barrier Sheeting and Sealers for Radon Control

One of the following products with low emissions may be specified for blocking entry of radon from the ground into the living space.

Barrier Sheeting and Sealers for Radon Control

- **Tu-Tuf 4**: Cross linked polyethylene sheeting.
- **Cross Tuff**: Specify radon control grade.
- **AFM Safecoat Dynoseal**: Waterproof, vapor proof, moisture proof, membrane sealer.

Chart 7-1: Resource List

Product	Description	Manufacturer/Distributor
100% Silicone Sealant	Clear sealant.	DAP/Dow Corning Products; 855 N. 3rd Street; Tip City, OH 45371; (800)634-8382. Available at many hardware chains including Home Depot, Ace Hardware, Hacienda Homecenters, and Builders Square
AFM Safecoat Caulking Compound	Water based elastic emulsion caulking compound designed to replace traditional caulk and putty for windows, cracks, and maintenance. Limited distribution.	AFM (American Formulating and Manufacturing); 350 West Ash Street, Suite 700; San Diego, CA 92101-3404; (800)239-0321, (619)239-0321 [Caulking Compound is also available through The Living Source catalog; P.O. Box 20155; Waco, TX 76702; (817)776-4878]
AFM Safecoat Dynoflex	Available in sprayable form to use as topcoat over Dynoseal.	Same
AFM Safecoat Dynoseal	Flexible, low odor, waterproof, vaporproof barrier.	Same
Air Krete	Cementitious foam insulation made of magnesium oxide, calcium, and silicate. R value equals 3.9/inch.	Nordic Builders; 162 N. Sierra Court; Gilbert, AZ 85234; (602)892-0603
Airchek Radon Land Test Kit	The open land sampler is a one-day test device mailed to the lab for reading.	Airchek Inc.; Naples, NC 28760; (800)247-2435
Brae Roof	Modified bitumen roofing.	For local certified applicators call US Intec Inc.; P.O. Box 2845; Port Arthur, TX 77643; (800)624-6832

Chart 7-1: Resource List

Product	Description	Manufacturer/Distributor
Celbar	Spray in or loose fill cellulose insulation treated with a borate compound as a fire retardant. Loose fill insulation can be ordered without recycled newspaper content.	International Cellulose Corporation; P.O. Box 450006; 12315 Robin Blvd.; Houston, TX 77245; (800)444-1252, (713)433-6701
CertainTeed	Manufacturer of undyed, unbacked fiberglass batt insulation.	CertainTeed Corporation; 750 East Swedesford; Valley Forge, PA 19489; (800)274-8530; For closest distributor, call (800)441-9850
Cord Caulk	Acrylic yarn saturated with adhesive wax polymers for sealing around doors, windows, and sills.	Delta Products Incorporated; 26 Arnold Road; North Quincy, MA 02171-3002; (617)471-7477, Fax: (617)773-4940 Available through the following mail order houses: Brookstone Company, (800)846-3000; Home Improvements, (800)642-2111; Alsto's Handy Helpers, (800)447-0048
Cross Tuff	Cross laminated polyethylene air barrier and under slab radon barrier.	Manufactured Plastics and Distribution Inc.; 2162 Market Street; Denver, CO 80205; (303)296-3516, (719)488-2143
Denny Foil Vapor Barrier	Virgin kraft paper with foil laminated to it on both sides with sodium silicate adhesive.	Denny Sales Corporation; 3500 Gateway Drive; Pompano Beach, FL 33069; (800)327-6616, (954)971-3100
GE 012	Clear silicone sealant.	GE; 260 Hudson River Road; Waterford, NY 12100; Technical service: (800)255-8886. Available through the following retail outlets: True Value, Ace Hardware, Home Base
GE 5091	Silicone paintable sealant.	Same
Greenwood Cotton Insulation	Batt insulation from recycled cotton. Handles like fiberglass, but safer to occupant and installer. R=3.7/inch.	Greenwood Cotton Insulation; P.O. Box 1017; Greenwood, SC 29648; (800)546-1332
Hi Tuff EP	Low odor, ethylene propylene, heat weldable roofing membrane.	Elastomerics Corporation; 9 Sullivan Road; Holyoke, MA 01040; (800)621-7663, (413)533-8100
Icynene Insulation System	Petroleum based sprayed on foam insulation. Good performance and extremely low outgassing make this product acceptable for many with chemical sensitivities.	Icynene Inc.; 376 Watline Avenue; Mississauga, ON Canada; (800)946-7325, (905)890-7325
Insulfoil	Air barrier of foil laminated on two sides of kraft paper with nontoxic adhesive.	Roy Akers Advanced Foil Systems; 4471 E. Santa Ana Street, Suite F; Ontario, CA 91761; (800)421-5947, (909)390-5125
Lithoseal Building Caulk	High quality urethane modified polymer. Inert once cured.	LM Scofield Company; P.O. Box 1525; Los Angeles, CA 90040; (800)800-9900, (213)723-5285

Chart 7-1: Resource List

Product	Description	Manufacturer/Distributor
Miraflex	Less toxic, undyed, fiberglass insulation material. The modified fibers are "safer," according to the manufacturer.	Owens Corning; Fiberglass Tower; Toledo, OH 43659; (800)438-7465
Mirrorseal	Nonpetroleum based polymer, fluid applied roofing system. Can be applied and repaired by unskilled labor without specialized tools.	Innovative Formulation; 670 W. 33rd Street; Tucson, AZ 85713; (800)346-7265, (602)628-1553
Phenoseal	Water based, nontoxic, nonflammable caulks and sealants available in translucent and 15 colors.	Glouster Company, Incorporated; P.O. Box 428; Franklin, MA 02038; (800)343-4963, (508)528-2200
Polyken Tape +337	Aluminum tape that forms an effective air barrier.	Kendall/Polyken; 15 Hampshire Street; Mansfield, MA 02048; (800)248-0147, (508)261-6200
Resource Conservation Technologies, Inc.	Acrylic polymer roll-on paint roofing without toxic dispersants or tints.	Resource Conservation Technologies, Inc.; 2633 N. Calvert Street; Baltimore, MD 21218; (410)366-1146
Silicone Plus	Paintable, water soluble, silicone sealant.	DAP/Dow Corning Products; 855 N. 3rd Street; Tip City, OH 45371; (800)634-8382; available at many hardware chains including Home Depot,Ace Hardware, Hacienda Homecenters, and Builders Square
Soil Gas Collector Matting	Used alongside perimeter at top of stem wall along with "T" risers and piping. Effectively removes radon gas before it enters building.	Professional Discount Supply; 1029 South Sierra Madre, Ste. B; Colorado Springs, CO 80903; (800)688-5776, (719)444-0646
Thermomaterials	Low VOC elastomeric roof coating.	U.S. Intec Inc.; 1594 Dawson Drive; Vista, CA 92083; (800)331-5228, (800)624-6832
Thoroseal Foundation Coating	Cementitious waterproofing for concrete surfaces.	Thoro Systems Products; 8570 Phillips Hwy., Suite 101; Jacksonville, FL 32256-8208; (800)322-7825, (904)828-4900
Tu-Tuf 3	High density, cross laminated polyethelene, puncture resistant air barrier.	Stocote Products Inc.; Drawer 310; Richmond, IL 60071; (800)435-2621, (815)675-6713
Tu-Tuf 4	Tu-Tuf 4 is thicker than Tu-Tuf 3 and can be effectively used under concrete.	Same
Volclay	4'x4' corrugated kraft panels filled with Bentonite clay that expand when wet to form a waterproof barrier.	Cetco; 1350 W. Shure Drive; Arlington Heights, IL 60004-1440; (800)948-5419, (708)392-5800

Further Reading

Lafavore, Michael. *Radon: The Invisible Threat.* Rodale Press, 1987.

Lstiburek, Joseph and John Carmody. *Moisture Control Handbook, Principles and Practices for Residential and Small Commercial Buildings.* Van Nostrand Reinhold, 1993.

U.S. Environmental Protection Agency. *A Citizen's Guide to Radon, Second Edition.* Washington, DC: U.S. Government Printing Office, EPA 402-K-92-001, May 1992.

U.S. Environmental Protection Agency. *Consumer's Guide to Radon Reduction.* Washington, DC: U.S. Government Printing Office, EPA 402-K-92-003, May 1992.

U.S. Environmental Protection Agency. *Indoor Radon and Radon Decay Reduction Measurement Device Protocols.* Washington, DC: U.S. Government Printing Office, EPA 402-R-92-004, July 1992.

U.S. Environmental Protection Agency. *Model Standards and Techniques for Control of Radon in New Residential Buildings.* Washington, DC: U.S. Government Printing Office, EPA 402-R-94-009, March 1994.

U.S. Environmental Protection Agency. *Radon Contractor Proficiency (RCP) Program.* Washington, DC: U.S. Government Printing Office, EPA 402-B-94-002, September 1994.

Division 8 - Openings

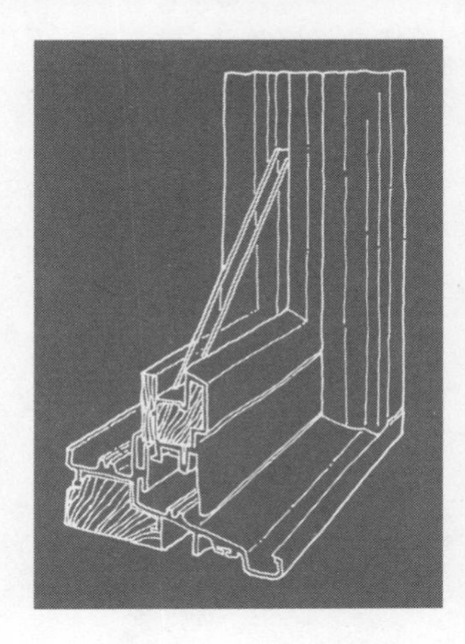

Wood Doors

Wood doors, both solid and paneled, are typically treated with biocides and manufactured with toxic glues. Paneled doors use less glue than solid veneered doors and will therefore outgas less. The face veneer commonly used on flush wood doors is Luan, made from mahogany trees which are stripped in the Philippines and Thailand in an operation damaging to the environment. Interior, fire rated doors often contain a particleboard core which will continuously offgas formaldehyde fumes. Factory made doors should be sealed to lock in harmful vapors. Recommended specifications appear below.

Factory Made Wood Doors

- All doors shall be thoroughly sealed on all six surfaces.
- For a clear finish, seal doors with one of the vapor barrier sealants listed in Division 9 - Finishes. Follow the manufacturer's instructions.
- For a painted finish, prime all six sides with **BIN Shellac**, and paint with one of the paints specified in "Division 9 - Finishes."

Choosing custom doors allows you the opportunity to select the type of wood and finishes. Some of the many custom door manufacturers will work with you to create a healthier product by using benign glues and less toxic shop finishes. You can also purchase the door unfinished. Athough most custom doors are more expensive, some custom door companies have production or builder lines which are almost cost competitive with manufactured doors. For custom doors, specify the following items.

Custom Wood Doors

- Doors shall be glued with a solvent free glue such as **Titebond Solvent Free Construction Adhesive**, **Elmer's Carpenter's Glue**, **AFM Almighty Adhesive**, or **Envirotec Health Guard Adhesive #2101**.
- Doors shall be finished by the contractor using the specified low toxic finish.

Sources for Custom Door Manufacturers

- Santa Fe Heritage Door Company: This company uses **Titebond** glues upon request and will either supply doors unfinished or work with your specified finishes.
- Spanish Pueblo Doors: Owner Will Ott uses solvent free glues and will work with the client to create custom finishes using low VOC products.

Window and Door Screens

Windows and sliding glass doors generally come with removable screens. Screen doors for French doors or glass swinging doors are not usually provided by the manufacturer and must be custom made. Aluminum screening was standard in the past, compared to fiberglass at present. Fiberglass is more flexible, more transparent, does not dent, and is easy to replace. Unfortunately, it can also be more odorous, especially if it has been treated. When selecting a window and door company, examine the screens for detectable odors. They will eventually air out, but if you find them objectionable, any custom screening company can replace them with aluminum screening.

Screens on crank-out casement or awning windows will have more impact on the indoor air quality because they are placed on the inside of the glass. Occupants will be exposed to these types of screens even when the windows are closed.

Windows

Wood Window Frames

Similar to most factory made doors, wood windows contain biocides and toxic glues and are a source of indoor air pollution unless they are sealed. Unlike wood doors, it is usually cost prohibitive to have operable windows custom made. Chemically sensitive individuals will often choose steel or aluminum windows with a baked-on enamel finish for this reason. Wood windows should be sealed to lock in harmful vapors. Specify the following items.

Wood Window Frames

- All windows shall be thoroughly sealed on all surfaces exposed to the interior with the use of a vapor barrier sealant specified in "Division 9 – Finishes."
- For a painted finish, prime with **BIN Shellac**, and paint with one of the paints specified in "Division 9 – Finishes."

Most wood window manufacturers also produce clad windows which contain wood on the inside and an aluminum, steel, or fiberglass coating on the outside for weather protection. This eliminates or severely reduces the need for exterior maintenance. If unclad windows are used, yearly maintenance is required with staining and sealing in order to protect them from the elements. The following products do not contain many of the toxic substances found in exterior wood finishing products. They can be more difficult to apply than standard products. Sampling a small area first may be advisable if the person applying the exterior finish has no previous experience with the products.

Exterior Wood Window Frames and Doors

The following products may be used for the exterior preservation of wood window frames and doors:

- **AFM Safecoat DuroStain**: Wood stain.
- **Auro No. 131 Natural Resin Oil Glaze**: Transparent, tintable finish.
- **Bio Shield Primer Oil #81**: Undercoat with finish coat of **Livos Kaldet Stain** or **Livos Vindo Enamel Paint**.
- **Bio Shield Transparent Wood Glaze**: Wood finish with ultraviolet protection.
- **OS/Color One Coat Only**: Weather and UV resistant, water repellant, semi-transparent wood stains.
- **WEATHER PRO**: Water based, water repellant wood stain.

Windows and doors in this ranch home were custom made in order to avoid the use of noxious glues and pesticide treated wood. (Architect: Paula Baker. Contractor: Living Structures, Inc.)

Weather Stripping

Weather stripping is used around doors to make them airtight and resistant to water leakage. Stripping can also be specified around interior doors where noise or odor control is desired. Many available weather strips are made of synthetics, including silicone, urethane foam, polypropylene nylon, and neoprene. Some will outgas. Neoprene, for example, can have a strong odor. Brass and stainless steel are also available at many hardware stores. Choose the least odorous weather stripping that accomplishes the job.

When planning a healthy home, the garage or mechanical room should not be designed to open to the interior of the home. Where a door to the garage or mechanical room opens into the living space, it is important to specify that these doors have a sealed threshhold, using a silicone or other acceptable sealant. The doors should be fully weather stripped in order to prevent harmful fumes from entering the living space. Consider the following specification.

Weather Stripping

All doors opening on to the living spaces from the garage or mechanical room shall be weather sealed. Provide a threshhold and seal with 100% silicone, aquarium grade.

Chart 8-1: Resource List

Product	Description	Manufacturer/Distributor
AFM Almighty Adhesive	Zero VOC, clamping adhesive for gluing wood and wood laminates. This product is available only through special order at a 50-gallon minimum.	AFM (American Formulating and Manufacturing); 350 West Ash Street, Suite 700; San Diego, CA 92101-3404; (800)239-0321, (619)239-0321
AFM Safecoat DuroStain	Seven different earth pigment, semi-transparent, wood stains. Interior/exterior, water based.	Same
Auro No. 131 Natural Resin Oil Glaze	Transparent, tintable finish for exterior porous wood protection. Note: This product is made with natural plant and mineral derivatives in a process called plant chemistry. Some sensitive individuals may have severe reactions to the natural turpenes, citrus derivatives, and oils.	Sinan Company; P.O. Box 857; Davis, CA 95617; (916)753-3104

Chart 8-1: Resource List

Product	Description	Manufacturer/Distributor
BIN Shellac	White shellac/sealer used to create an effective air barrier.	Wm. Zinsser & Company; 39 Belmont Drive; Sommerset, NJ 08875; (732)469-8100
Bio Shield Penetrating Oil Primer #81	Sealer, undercoat, and primer for absorbent surfaces of wood, cork, stone, slate, and brick.	Eco Design/Natural Choice; 1365 Rufina Circle; Santa Fe, NM 87505; (800)621-2591, (505)438-3448
Bio Shield Transparent Wood Glaze	Interior/exterior wood finish with ultraviolet protection.	Same
Elmer's Carpenter's Glue	Solvent free glue.	Borden Inc.; 180 Broad Street; Columbus, OH 43215; (800)426-7336, (800)848-9400. Available in many retail outlets.
Envirotec Health Guard Adhesive #2101	Zero VOC, solvent free adhesive without alcohol, glycol, ammonia, or carcinogens.	WF Taylor Company Inc.; 11545 Pacific Avenue; Fontana, CA 92337; (800)397-4583, (909)360-6677
Livos Kaldet Stain	Stain and finish oil in 12 colors for interior and exterior surfaces made of wood, clay, or stone. Note: Livos products are made from plant and mineral derivatives. As with many natural products, some chemically sensitive individuals may not tolerate turpenes, oils, and citrus based or other aromatic components found in these formulations.	Eco Design/Natural Choice; 1365 Rufina Circle; Santa Fe, NM 87505; (800)621-2591, (505)438-3448
Livos Vindo Enamel Paint	Wood finish coat.	Same
OS/Color One Coat Only	Twelve different stain colors in base of vegetable oils. Interior/exterior use. No preservatives or biocides.	Ostermann & Scheiwe; P.O. Box 669; Spanaway, WA 98387; (800)344-9663
Santa Fe Heritage Door Company	Custom wood doors.	Santa Fe Heritage Door Company; P.O. Box 2517; Santa Fe, NM 87501; (505)473-0464
Spanish Pueblo Doors	Custom wood doors and cabinets.	Spanish Pueblo Doors; P.O. Box 2517; Santa Fe, NM 87501; (505)473-0464
Titebond Solvent Free Construction Adhesive	Solvent free adhesive.	Franklin International; 2020 Bruck Street; Columbus, OH 43207; (800)347-4583
WEATHER PRO	Water based, water repellant wood stain for interior/exterior. Volatile organic compound (VOC) compliant.	Okon Inc.; 6000 W. 13th Avenue; Lakewood, CO 80214; (800)237-0565, (303)232-3571

DIVISION 9 - FINISHES

Finishes include all surface materials and treatments in the home. Finishes comprise what is seen on a daily basis, and constitute the personal signature of the owner. Finishes are the predominant source of smells in a new home. They can introduce a multitude of volatile organic compounds into the air and will continue to volatilize, or outgas, for years after the home is completed. However, when chosen carefully, finishes can enhance health and well-being as well as add to the aesthetic value of the home. Until recently, nonpolluting finishing products were considered specialty items.

Fortunately, healthier solutions are now regularly appearing on the market. Many of these items are easily accessible, cost competitive, and comparable in performance to their more toxic counterparts. Some even have the ability to seal in toxins that may be present in underlying materials, thereby improving air quality.

In some regions there is widespread use and availability of traditional nontoxic finish materials. For example, in the Southwest, tiles, stones, stuccos, and plasters are commonplace, whereas in many regions of the country they have been replaced by vinyl, laminates, and other synthetic substitutes, even in custom homes.

In building a healthy home, we would encourage you to take full advantage of the traditional materials native to your region.

Case Study: Immune dysfunction related to formaldehyde exposure in the home

P.F. is a 43-year-old woman who was in good health until 1981 when she moved into a new mobile home. Shortly thereafter she developed a digestive disorder with gas and bloating, severe insomnia, and a chronic cough with frequent episodes of bronchitis. By the following year she was suffering from persistent fatigue and frequent respiratory infections, including her first case of pneumonia. She became sensitive to most products containing formaldehyde, especially pressboard. She noted that she experienced "brain fog" while shopping at the local mall. Her symptoms continued to worsen, and now included allergies, hypoglycemia, and lethargy.

P.F. consulted with several healthcare practitioners, including a pulmonary specialist, psychiatrist, hypnotist, nutritionist, acupuncturist, and many more. None of them ever questioned her about the air quality in her home. Eventually she received the diagnosis of multiple chemical sensitivity from a physician with similar symptoms, and was finally educated as to the underlying cause of her health problems. In 1992, P.F. moved into a house that contained low formaldehyde levels which alleviated some of her symptoms. Her house contained several healthful features such as radiant heat in concrete floors, and the absence of pressboard and particle board in its construction.

However, further modifications were necessary before her health could be stabilized and improved. All gas appliances were removed; filtration was installed for both air and water; and the mechanical room was vented to the outside. By 1996 P.F. had regained her health. However, as is typical in such cases, she still becomes symptomatic upon re-exposure to toxic fumes and must diligently maintain a "safe" environment for herself.

Discussion

Indoor formaldehyde is gaining recognition as a severe health hazard to occupants of homes and office buildings where chronic exposure occurs. Several organizations, such as the American Lung Association, have recommended that formaldehyde levels not exceed 0.1 parts per million. People who have already become sensitized to formaldehyde will have reactions at levels as low as 0.02 ppm. Approximately 50 percent of the population is exposed on a daily basis in the workplace to levels which exceed the 0.1 ppm limit. Mobile homes are notorious for causing health problems related to extremely high levels of formaldehyde emitted from the plywood and particle board used in construction.[1]

Individuals who develop permanent health problems associated with formaldehyde exposure often relate the onset of their symptoms to a flu-like illness which is diagnosed as a viral infection. However, the affected individual usually does not totally recover from this so-called flu and is left with general malaise, fatigue, and depression. Other symptoms can include rashes, eye irritation, frequent sore throats, hoarse voice, repeated sinus infections, nasal congestion, chronic cough, chest pains, palpitations, muscle spasms and joint pains, numbness and tingling of the extremities, colitis and other digestive disorders, severe headaches, dizziness, loss of memory, inability to recall words and names, and disorientation. Formaldehyde is an immune system sensitizer, which means that chronic exposure can lead to multiple allergies and sensitivities to substances that are entirely unrelated to formaldehyde. This is known as the "spreading phenomenon."

P.F. was typical of people whose multiple chemical sensitivities stem from formaldehyde exposure in that she consulted numerous physicians and specialists in an attempt to obtain a diagnosis for her chronic ill health. Physical examinations and standard testing usually fail to identify the cause of such health problems. Sometimes it is suggested that the patient is a hypochondriac or in need of psychiatric evaluation. When asked if there might be a connection between the symptoms and formaldehyde, most physicians either do not know or are of the opinion that formaldehyde merely causes irritation. As a result, the patient's health continues to deteriorate due to continued exposure.

Plaster and Gypsum Board

Plaster

Plaster generally provides a healthful interior wall finish. Because of the labor and skill involved in its application, it is a more expensive finish. Because of its beauty, it is much sought after. Plaster has the ability to block VOCs present in small quantities in the gypsum lath and taped joints which comprise its base in frame construction. In pumicecrete, straw/clay, and adobe construction, the plaster may be applied directly to the wall material. Although most plasters are inert, some plasters contain polyvinyl additives subject to outgassing and should be avoided. Any additives in plaster should be clearly marked on the packaging. Nevertheless, it is best to verify additives with the manufacturer prior to purchase.

One potential health hazard associated with plaster lies in the method in which it is dried. Because new plaster releases a significant amount of moisture, drying it out quickly becomes necessary before other building materials are adversely affected. This is especially problematic in the winter months when the cold temperatures and lack of ventilation slow down the rate of evaporation. The standard solution is to use gasoline or kerosene heaters. The by-products of combustion generated by this machinery are readily absorbed into the plaster and other building materials, and therefore we do not recommend this practice. Electric heaters tend to be more expensive to run with far less BTU output. We recommend a combination of careful scheduling so that the plaster work occurs in a warm dry period, and the use of electric turbo high velocity dryers and dehumidifiers when necessary.

Plaster Installation

- Plaster shall be free of additives.
- The use of gas or kerosene generated heaters within the building envelope is prohibited.
- Turbo high velocity heaters, other electric heaters, electric dehumidifiers and blow-in heaters with respective combustion sources outside the building envelope are acceptable.

Plaster Finish

Because of the porous nature of plaster, it will stain and show fingerprints if left unfinished. Plaster walls, which were the norm before the advent of plasterboard or sheetrock, were commonly painted or covered with wallpaper. Recently plaster has become a more prestigious finish material. Most people prefer to apply a clear finish over it to protect and enhance its natural beauty.

Natural beeswax finishes will protect the wall while maintaining its "breathability." Traditionally, beeswax was applied with a hot knife and troweled on the wall. There are very few craftsmen who know this art form today. However, we have found a healthful beeswax furniture polish which can simply be applied with a cloth and buffed. Some synthetic finishes will create a more impervious seal and are less expensive, easier to apply, and more enduring. Synthetic finishes should be carefully evaluated for suitability. Some may be toxic or increase problems with static electricity. Since they reduce breatheability of the plaster, they may encourage mold growth if moisture becomes trapped. We have successfully used the finishes in the specifications below.

Plaster Finishes

- **Livos Glievo Liquid Wax**: Apply a thin coat and hand buff.
- **Okon Seal and Finish**: For satin gloss.

Gypsum Board

Gypsum board, also known as gypboard, sheet rock, or drywall is the most common form of interior wall sheathing in residential construction. It is considerably less expensive than plaster. The 4'x8' sheets are attached to the studs then taped, sealed, textured, and painted.

Gypboard is composed of natural gypsum sandwiched between two sheets of cardboard that are made from recycled newsprint. It can be purchased with aluminum foil backing on one side. While the foil acts as a barrier against outgassing materials within the wall cavity, it may also lock in moisture resulting in microbial growth. For that reason we do not recommend the foil backing.

The installation of gypboard in standard practice may negatively impact indoor air quality. Five of the most important characteristics that lead to poor air quality follow.

1. Dust and debris within wall cavities are often enclosed and concealed by the gypboard. If dust and debris are not cleaned out, they can cause problems over time. Dust can eventually work its way back into the living space and become a maintenance problem as well as an air pollutant.
2. The gypboard itself will outgas because of the inks remaining in the recycled newspaper.
3. The standard premixed joint compounds contain several undesirable chemicals, including formaldehyde.
4. Like plaster, gypboard is highly absorbent. In standard practice, gas and kerosene heaters are used to dry the joint compounds. The byproducts of combustion are absorbed into the walls and will outgas into the building envelope of the completed home.
5. The cardboard on gypsum board can act as a nutrient for mold growth if it gets wet. If the gypsum board sets flush with the floor, water can easily be "wicked up" the wall during an accidental leak or spill, or minor flooding. To prevent wicking, set the gypboard 1/2" above the floor.

Gypboard Installation

- All wall cavities shall be thoroughly vacuumed and free of debris prior to installation of the gypboard.
- Joint compound shall be Murco M-100, a powdered joint cement and texture compound formulated with inert fillers and no preservatives.
- Heaters fueled by gasoline or kerosene are prohibited.
- The joint compound must be completely dry before the application of primer.
- In order to seal in VOCs that are generated by the surfacing board, all gypboard walls must be primed with one of the following prior to painting: (1) **BIN Primer Sealer,** a white shellac primer that is a vapor barrier sealer; (2) **AFM Safecoat New Wallboard Primecoat HPV**, a water reducible primer; or (3) **86001 Seal**, a clear, water reducible, primer sealer.
- In plumbed areas such as bathrooms, kitchens and mechanical rooms, and in walls adjacent to these rooms, set gypboard with a 1/2" minimum gap above the finished floor and provide a baseboard or tile trim.

Tile

Tile is generally an inert and healthful floor, wall, and counter surfacing material. However, the following concerns must be addressed in order to achieve a healthful installation.

- In standard construction, tile is often laid over an unacceptable backing such as particleboard or greenboard in wet areas.
- Certain imported tiles contain lead based glazes or asbestos fillers.
- Certain glazes, primarily imports, have been found to be radioactive, especially cobalt blues and burnt oranges.
- Prefinished, factory glazed tiles are preferable to unglazed tiles.
- Many tile sealing products contain harmful chemicals with high levels of VOCs.
- Most standard tile adhesives and mortars contain harmful chemicals and should be avoided.
- Standard grouts usually contain fungicides and latex additives.
- Grouts are porous and can harbor mold and mildew. They should be sealed where exposed to water.

Tile Setting Materials and Accessories

Underlayment for Ceramic Tile

The following are acceptable underlayments for ceramic tile:

- A clean, level, concrete slab or gypcrete.
- Exterior grade plywood. Use only where a cementitious underlayment is unavailable. This method will require mastic adhesive.
- **Medex**: A nonstructural, formaldehyde free, medium density fiberboard. Tile application will require mastic adhesive. The board is not waterproof and will either require a waterproof sealer such as Safecoat Safe Seal or sufficient time for the board to dry out completely prior to application of grout.
- **Cemroc**: A lightweight, strong, noncombustible, highly water resistant cementitious board.
- **Hardibacker Board**: A cementitious tile backer board for backing tile installations in wet locations.

Tile Installation

The three basic methods for installing tiles are thickset, thinset, and organic mastics. Descriptions of the methods are covered in the following sections.

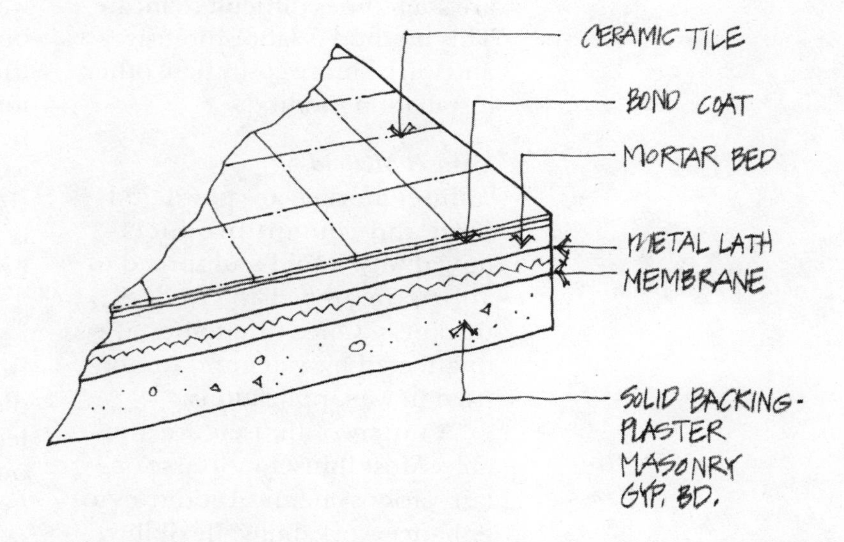

Ceramic tile installation.

Thickset Method

This is the tried and true, old fashioned way of adhering tiles, prevalent prior to the invention of additives. A thick reinforced bed of mortar consisting of Portland Cement sand, and in some cases lime, is floated. While the base is still plastic, a thin layer of Portland Cement paste, known as the bond coat, is spread over it and the tile is adhered to the bond coat, allowed to cure for several days, and then grouted.

This method will create the strongest, most durable tile installation available without the use of chemical additives. The following should be specified for this type of installation.

Thickset Tile Installation

- Use only additive free Portland Cement, clean sand, lime where required, and potable water. Use the recommended reinforcing and specified cleavage membrane.
- For walls, ceiling, and floor installations, follow the method covered by ANSI AI08.1, and set the tiles on the mortar bed while it is still plastic.
- The cleavage membrane shall be non-asphalt impregnated, 4 mm polyethylene such as **Tu-Tuff** or an approved equivalent.

It is not always possible to use a thickset installation. The 1-1/4" depth required for thick setting may not be available unless carefully planned from the outset. Tile setters skilled in this method are sometimes difficult to locate. This method is labor-intensive and will be more costly than other installation methods.

Thinset Method

Thinset mortars are powdered sand and cement products mixed with liquid and spread to approximately 1/8" to 3/8" thickness. Once dried, they are unaffected by water and can be used in wet applications.

A variety of thinsets are available. Most thinset mortars contain various chemical additives to enhance workability, flexibility, and bonding strength, thus expanding the range of application. Water mixed thinsets consist of powdered sand and cement. They are available with or without powdered latex and acrylic additives, and are mixed on site with water. Latex and acrylic thinsets consist of powdered sand and cement mixed with liquid latexes and acrylics instead of water. They have higher bond and compressive strength and improved flexibility compared to thinsets mixed with water. Epoxy thinsets develop bonds more quickly than other thinsets. The epoxies emit noxious fumes while curing. Protective vapor respirators should be worn during application. The risk these fumes pose to workers is almost always unnecessary in home construction.

The actual additive ingredients used in these thinset mixtures are proprietary and not disclosed on the label. When selecting a thinset product, choose one that has the smallest amount of chemical additives necessary to do the job at hand and choose the brand with the least odor.

The following water mixed thinsets are available without synthetic additives. They may be used successfully over clean concrete slabs, properly supported cementitious boards, and mortar beds.

Thinsets Without Synthetic Additives

- **Laticrete Additive Free Thinset**
- **C-Cure Floor Mix 900**, **Wall Mix 901**, and **Thinset 911** (dual purpose)

The previous two products constitute the "economy" line of C-Cure Mortars which contain no additives other than mineral salts.

The following water mixed thinsets contain vinyl polymers which give them greater strength and range of application. They have very little odor and are virtually odorless once cured. Anyone with sensitivities to vinyl polymer additives, however, is advised to test these products prior to using them.

Low Odor Thinsets with Vinyl Polymer Additives

- **C-Cure Permabond 902**: A dry set mortar with Portland cement, sand, and additives for use over cementitious substrates.
- **C-Cure Multi-Cure**: A latex enhanced dry set mortar with added bonding strength and flexibility for use over cementitious and plywood substrates.

Organic Mastics

Organic mastics are either latex or petrochemical based adhesives that consist of a bonding agent and a liquid vehicle. For petroleum based mastics, the vehicle is a solvent, usually toluene. These formulations are both highly toxic and flammable, and are not recommended for use in a healthy home. The vehicle in latex based mastics is water.

Organic mastics enjoy widespread popularity because they are inexpensive, stickier than thinsets, and allow the quickest installations. However, they do not have the strength, flexibility, or water resistance of thinset or thickset applications. Because they are applied very thinly, they do not have leveling capabilities and they are only suitable for application over flat surfaces such as plywood or drywall. Most mastics are not recommended for areas that get wet. Where mastic applications are appropriate, specify the following.

Organic Mastics

Mastics containing solvents are prohibited.

The following water soluble adhesives may be used where an organic mastic application is appropriate:

- **AFM Safecoat 3 in 1 Adhesive**: Low odor, low VOC, water based mastic for hard composition wall and floor tiles.
- **Envirobond #801**: Water based latex mastic which can be used in wet areas. The product is VOC compliant, containing no toluene, hexane, or benzene. Allow it to cure prior to grouting.

Tile slate and glass combine to create this shower enclosure which opens on to beautiful surrounding views. (Architect: Paula Baker. Contractor: Prull & Associates, Inc. Photo: Lisl Dennis.)

Grouts

As with tile setting mortars, there are a number of additives that may be used in commercial grouts to impart certain performance characteristics such as improved strength and flexibility, increased water or stain resistance, and improved freeze-thaw stability. Some of these additives, such as epoxies, are quite noxious. In grout applications, they will be exposed to the living space and will continue to outgas until completely cured. Grouts can be mixed on site by combining Portland cement, sand, lime (optional), and water. They can be colored with the same pigments as used to color concrete. (Refer to "Division 3 – Concrete.") It is important that the person mixing the grout be knowledgeable as to proper proportions and sand size for the particular tile application. These applications should be damp cured for three days. The following commercially available grouts are free of latex additives.

Additive Free Grouts

- **C-Cure AR Grout**: A sanded grout available in a wide selection of colors.
- **C-Cure Supreme 925 Grout**: An unsanded grout for joints less than 1/8" and for use with tiles that are easily scratched, such as marble.
- **Ceramic Tile Grouts**: Sanded or unsanded, available in 35 colors.
- **Mapei 2-1/2 to 1**: For large grout joints greater than 3/8".

Grout Sealers

Sealing grouts will make them easier to clean and more resistant to water penetration and staining. Water penetration of grout joints makes them susceptible to mold and bacteria growth. We do not recommend the commercially available grouts enhanced with additives. The following products may be specified as grout sealers.

Grout Sealers

- **AFM Safecoat Safe Seal**: An odorless, zero VOC, water based, low gloss sealer for highly porous surfaces. Can be diluted in a 50:50 ratio with water, then mixed into the dry grout to form an integral grout sealer.
- **Sodium Silicate**: Also known as water glass. A clear liquid sealer that can be painted over grout joints.

Tile Sealers

There is a very wide variety of factory glazed tiles from which to choose. Many of these tiles have a commercial rating and will never require refinishing. They are preferred for the healthy home.

If an unsealed tile is selected, then it is important to specify a healthful sealant. Many of the commercially available tile sealers are solvent based, highly toxic products that will emit noxious fumes for a long time after application. Consider specifying the following products.

Tile Sealers

- **Pace Crystal Shield**: Odorless sealer once dry; available in velvet or gloss finish.
- **AFM Safecoat MexeSeal** over **AFM Safecoat Paver Seal .003**: For sealing previously unsealed tile floors.
- **ZipGuard Environmental Water Base Urethane**: Can be used on very clean, previously unsealed tile floors for sealing.

Stone

While stone is generally a healthful and beautiful choice for flooring and decorative accents, the concerns for proper installation are the same as for ceramic tile. The specifications we have outlined for ceramic tile also apply to stone. We have tested several stone products for radiation and radon content and found a range of readings from very low to high levels. Although uranium content in construction materials is not usually considered to be a serious concern, John's experience, as shared in the following case study, leads us to conclude that stone can contribute significantly to ambient radon levels in a home. We recommend that stone, especially granite, be screened for radon prior to installation, even though the Granite Institute has issued a scientific report concluding that granite counter tops do not emit radioactivity into the home. Measurements are easily performed as described in "Division 13 - Special Construction."

Case Study:
A very hot bed

Prior to purchasing a home, a family contacted a consultantcy to conduct radon testing with electronic monitors, following the EPA's protocol. Closed house conditions were established 12 hours prior to beginning testing and were maintained throughout the tests. During the testing one of the electronic monitors located in the dining room indicated 12.5 pico-curies of radon per liter of air, while another monitor elsewhere showed close to normal levels. The client was advised that the electronic readings were suspicious and that additional testing was necessary. As the investigation proceeded, it became clear that there was a radon source at one end of the home. In fact, the radon results for a test conducted on a night table in the guest bedroom was 27.0 pico-curies, while the family room a short way down the hall was 7.0 pico-curies. The further the monitors had been placed from the guest bedroom, the lower the radon value.

Upon visual examination of the guest bedroom, it was noted that the headboards for the two beds were made of rock that appeared to be granite. The headboards were later measured with a small Geiger counter. While normal radioactive background levels away from the headboards were approximately 12 radioactive counts per minute, the counts close to the headboards were over 300. It was clear that the headboards were at least one source of radon in the room.

The headboards were in fact a decorative granite rock imported from Italy. Each headboard weighed several hundred pounds. The floors and walls had been especially constructed to hold the extra weight. It took six strong men to remove each of the headboards to a detached garage. The radon tests were repeated throughout the home with all values now under 1.0 pico-curie. The home was given a radon clearance, contingent upon the proper disposal of the headboards.

This was the first home John inspected in which a radiation source was caused by a building material or furnishing. Although radon from the soil is the most common cause of elevated radiation levels in a home, there are many other possible sources. Since granite rock is sometimes high in uranium, it must be considered a potential source of radon when used in construction. Rock can be a superb building material, but it should always be tested prior to use for the rare possibility of radiation.

Stone Installation

Refer to "Thickset Method" in the "Tile Installation" section. Refer also to the section on stone countertops in "Division 6 - Woods and Plastics."

Sealers for Stone

The following finishes are recommended for stone flooring, shelving, and countertops other than granite.

- **Livos Bilo Floor Wax**: A clear, mar resistant finish wax.
- **AFM Safecoat MexeSeal**: A durable sealer providing water and oil repellency, applied over **AFM Safecoat Paver Seal .003**, an undersealer for porous materials.
- **Naturel Cleaner and Sealer**: Water soluble flakes that clean, protect, and finish stone surfaces.
- **Livos Meldos Hard Oil**: A penetrating oil sealer.

Flooring

Wood Flooring

In standard construction, wood floors are commonly laid over formaldehyde emitting underlayment, then finished with solvent based finishes which will outgas for many months. Wood is a healthy choice for flooring provided that the subflooring, adhesives (if used), and finishes are carefully chosen to be healthful as well. There are several premanufactured wood veneer flooring systems available. Each system must be carefully analyzed in terms of all individual components, from underlayment to finish. We recommend that the following be specified.

Wood Floor Installation

Underlayment for Wood Flooring

Interior grade plywood is prohibited.

The following underlayments are acceptable:

- 1" or 2" tongue and groove wood or rough sawn lumber laid diagonally.
- Exterior grade plywood (CDX), if used for underlayment, should be sealed with an acceptable vapor barrier sealant, as specified in "Division 9 - Finishes."
- Use only solvent free adhesives or 100% silicone. Refer to the adhesives section in "Division 6 - Wood and Plastics."

Sealers and Waxes for Wood Flooring

- **AFM Safecoat Hard Seal**: For medium gloss.
- **AFM Safecoat Polyureseal** or **Polyureseal BP** over **AFM Safecoat Lock-In New Wood Sanding Sealer**.
- **Bio Shield Penetrating Oil Sealer #5** by itself or as an undercoat with **Bio Shield Hard Oil #9**, or as an undercoat with a topcoat of **Bio Shield Natural Resin Floor Finish #92**.
- **Livos Ardvos Wood Oil** or **Livos Meldos Hard Oil**: Medium to high gloss.
- **Livos Bilo Floor Wax** or **Livos Glievo Liquid Wax**: Plant chemistry or beeswax products.
- **OS/Color Hard Wax/Oil**: A satin matt oil or wax finish.
- **Pace Crystal Shield**: Clear, durable seal for hardwood floors.
- **Skanvahr**: A very low odor, waterborne, clear, nonyellowing, self-cross linking finish.
- **Zip Guard Environmental Water Base Urethane**. Clear finish.

Resilient Flooring

Easy cleanup, economy, and a soft walking surface have made sheet vinyl a popular flooring for kitchen and utility areas. Yet vinyl flooring is associated with health hazards. Vinyl chloride fumes emitted from the vinyl flooring are a known carcinogen. In addition, in hot or humid climates requiring air conditioning, the vinyl will trap moisture which can promote delamination and mold growth or rot. We do not recommend vinyl in the healthy home. Natural linoleum made from linseed oil, pine resins, wood powder, and jute is free of toxic chemicals. Natural linseed oil does, however, have a noticeable odor associated with it which some people with extreme sensitivities do not tolerate. Cork tile is another healthy choice for flooring. Natural linoleum and cork can be used in hot, humid climates, providing the material is allowed to breathe. This means avoiding adhesives that will impede the vapor permeability of the materials. An acceptable finish for this flooring is beeswax.

Sources for Natural Cork and Linoleum Flooring

- **Bangor Cork Company**: Cork tiles and sheet flooring and linoleum.
- **DLW Linoleums**: Natural linoleums in a variety of colors with natural jute backing.
- **Eco Design/Natural Choice**: Cork floor tiles and adhesives.
- **Forbo Industries**: Natural linoleum and cork flooring products.
- **Hendricksen Naturlich**: Cork and other natural floor coverings and adhesives.
- **Natural Cork C° Ltd.**: Offers cork in a variety of colors, patterns, and finishes.

Adhesives for Natural Cork and Linoleum Flooring

- **AFM Safecoat 3 in 1 Adhesive**
- **Auro No. 383 Natural Linoleum Glue**
- **Bio Shield Cork Adhesive**
- **Envirotec Health Guard Adhesive #2027**

Case Study: Toddler made severely ill by carpet

B.J. is a two-year-old boy who was in excellent health until the age of 10 months at which time he suddenly developed seizures. These episodes of rigidity and tremors occurred up to 40 to 50 times a day. The baby was subjected to a series of invasive diagnostic evaluations by many different specialists. The blood tests, brain scans, and electroencephalograms revealed no apparent cause of the seizures. The baby was placed on medication to suppress the central nervous system. The seizures persisted although their intensity declined.

The baby's grandfather, a building contractor, suggested that the culprit might be the expensive new carpet installed shortly before the onset of the seizures. The parents contacted a representative from the carpet industry who denied any similar complaints of neurological problems from customers. The parents suspected that this information was incorrect. They sent samples of the carpet to the independent Anderson Labs in Vermont for testing. Air was blown across the carpet samples into the cages of mice whose symptoms were then observed and documented. After a short period of time elapsed, the mice developed tremors, rigidity, and seizures. The parents were horrified by the report. It was clear that their beautiful, new carpet had essentially poisoned their son. The carpet and pad were immediately removed from the home, the adhesive scraped off, and the house aired out. The seizures stopped. The child is now off all medication and doing much better, although blood testing shows immune system damage consistent with chemical injury.

EPA Takes a "Stand" on the Carpet Controversy

In October, 1987, the EPA began carpet installation at its headquarters in Washington, DC in the Waterside Mall. A total of 1,141 complaints were received regarding adverse health effects related to the new carpet.[2] These complaints included decreased short-term memory, loss of concentration, confusion, anxiety, headaches, joint and muscle pains, rashes, digestive disorders, reproductive abnormalities, asthma, insomnia, chronic fatigue, and multiple chemical sensitivities. Dozens of workers remained permanently disabled. After the EPA investigated these carpet complaints from its headquarters building, it published a report showing a positive correlation between the EPA worker complaints and the new carpet.[3]

Despite the results of its own study, and the removal of 27,000 square yards of carpet from the headquarters building in 1989, the EPA published a public information brochure which states, "Limited research to date has found no links between adverse health effects and the levels of chemicals emitted by new carpet."[4]

EPA's Director of Health and Safety told the *Washington Times* that, "the freshly manufactured carpet clearly caused the initial illness." Within a few weeks of making that statement he was removed from his job. EPA management expressed concern that testing and regulation of carpet emissions could potentially cost the carpet industry billions of dollars.[5]

Discussion

The Consumer Product Safety Commission (CPSC) has received hundreds of complaints about carpets causing respiratory and neurological problems.[6] Toxic emissions from carpets include fumes from formaldehyde, benzene, xylene, toluene, butadiene, styrene, and 4-phenyl-cyclo-hexene (4PC). These chemicals can potentially cause cancer, birth defects, reproductive disorders, respiratory problems, and neurological damage such as anxiety, depression, inability to concentrate, confusion, short-term memory loss, and seizures. The carpet industry has consistently denied adverse health effects of carpeting in spite of overwhelming evidence to the contrary.

In 1992, in response to public concern, the carpet industry announced its Green Tag program which has lured consumers into a false sense of safety. The program tests only a small sampling of carpets once a year. The testing is based only on volatile organic compound emissions, not biological health effects.[7] In fact, some carpets from the Green Tag program tested at the Anderson Labs have caused death to the mice exposed to carpet fumes.[8]

Carpeting

Carpeting has been associated with a growing number of health problems. In a typical carpet toxic chemicals may be found in the fiber bonding material, dyes, backing glues, fire retardant, latex binder, fungicide, and anti-static and stain resistant treatments. In 1992 during a congressional hearing on the potential risk of carpets, the U.S. Environmental Protection Agency (EPA) stated that a typical carpet sample contains at least 120 chemicals, many of which are known to be neurotoxic. Outgassing from new carpeting can persist at significantly high levels for up to three years after installation. Once discarded, carpet is neither renewable nor biodegradable. In major cities, discarded carpeting accounts for 7% of the landfill mass!

The most common carpet backing, synthetic latex, contains approximately 100 different gases which contribute to the unpleasant and harmful "new carpet smell." Most underpads are made of foamed plastic or synthetic rubber and contain petroleum products which cause pollution in every stage of production and continue to pollute once installed. Felt type backings are generally less polluting. We have specified safer carpet backings below. Typically, brands labeled "hypo-allergenic" will be odorless.

There are two ways to install wall to wall carpeting: tack down or glue down. Tack down installations are preferable because they are easier to remove, they do not destroy the floor surface, and the carpeting can be partially recycled. Tacking strips are nailed, screwed, or glued down around the perimeter of the room. It is important that the tacking strips, if glued, are attached with a low toxic glue. The carpet and underpad are then stretched, and the edges are folded with the underside tacked down.

Most standard adhesives for carpet installation are solvent based and contain harmful chemicals. Where a glue-down installation is required, solvent based adhesives are to be avoided. We have specified several healthier options below. In either installation procedure, seaming tapes will be required to fasten sections of carpeting together. Safer seaming tapes are specified below.

There are several untreated natural fibers available for wall to wall installations including wool, coir, and sisal. When these are installed with low or nontoxic backing and either tack down or low toxic glue installation, they will provide a safer solution than most standard installations. (See sample carpet installation specifications below.)

WARNING: Wool carpets are often treated with highly toxic mothproofing pesticides. Therefore, an expensive 100% wool carpet does not necessarily mean a safer carpet.

Wall to wall carpeting, whether standard or natural, serves as a reservoir for dirt, dust, mold, bacterial growth, and toxins tracked in from outside, even when regularly vacuumed and shampooed. Typical cleaning agents for wall to wall carpets contain harmful ingredients, including perfumes, chemical soil removers, brighteners, and antibacterial agents.

In new construction, home owners are typically given an allowance and asked to choose the carpeting. This allowance can also be used towards the purchase of healthier floor coverings.

Although we strongly recommend the use of throw rugs of natural fibers which can be removed and cleaned instead of wall to wall carpeting, we offer the following guidelines for selecting the least toxic carpeting for those who choose to use it.

- Verify with manufacturer that wool carpets have not been mothproofed.
- Of the synthetic carpets, 100% nylon is considered to be one of the safest.
- Choose carpeting which has little or no odor. Even the slightest odor on a small sample will be magnified many times in a fully carpeted room and can result in a very prominent, unpleasant and unhealthy smell.
- Choose your carpeting as early as possible so it will have the most time to air out prior to installation.
- Buy carpeting from a supplier who will agree to warehouse the carpet for you. This means that the carpet will be unrolled and aired out in the warehouse prior to shipping.
- Avoid carpeting that contains antimicrobial agents such as fungicides and mildewcides.
- Avoid carpeting containing permanent stain resistance treatment.
- Avoid carpeting or pads containing styrene-butadiene rubber.
- Carpeting with woven backing is preferable to rubberized backing.

- Follow underpad and installation recommendations in these specifications.
- Use nontoxic and odor free shampoos, and maintain carpets regularly to prevent mold, bacteria, dust, and pesticide buildup.
- Vacuum the carpets on a regular basis, moving furniture if necessary, to reach all areas where larvae may hide in order to prevent moth infestations in untreated wool carpets. A true HEPA (High Efficiency Particulate Accumulator) type vacuum cleaner is a must if you have carpet. It is the only type that collects the very tiny particles like dust mite feces and mold spores.
- Establish a no shoes policy for your home.
- If the carpet or pad gets wet, dry it as quickly as possible to prevent microbial growth.

WARNING: Never use wall-to-wall carpet in bathrooms, kitchens, laundry rooms, or mechanical rooms because carpeting in these areas inevitably becomes damp, inviting mold and bacteria infestation.

The following products and treatments are acceptable for carpet installation and may be specified.

Carpet Installation

For tack down installations where baseboard is used, terminate the carpet at the trim. Do not run the trim over the carpet.

Sources for Nontoxic Underpadding

- **Endurance II**: Synthetic Jutepad in 20 or 32 ounce weights.
- **Hartex Carpet Cushion**: Available in three weights.
- **Hendricksen Naturlich**: Recycled felt underpadding, heat bonded with no chemical additives.

Adhesives and Seaming Tapes for Carpet Installation

- **AFM Safecoat Almighty Adhesive**: Available by special order only.
- **Auro No 385 Natural Carpet Glue**
- **Hendricksen Naturlich Manufacturer's Adhesive**
- **Envirotec Health Guard Adhesive #2027, 2045, 2054, 2055, 2060, 2070, 2080**
- **Envirotec Health Guard Seaming Tapes #3070, 3080, 3090, 3093, 3094**

Carpet Treatment

The following carpet treatment will help remove pesticides, formaldehyde, and other chemicals from the carpeting and pad, and will also seal in chemicals to prevent outgassing. The treatment is not suitable for carpets with a large wool content.

- Treat carpet with **AFM SafeChoice Carpet Shampoo**, **AFM SafeChoice Carpet Guard**, and **AFM SafeChoice Lock-Out**.
- Follow manufacturer's instructions.
- Test a small sample of carpet with these products for shrinkage and color fastness prior to full application.

Home interior featuring nontoxic finishes and materials including brick flooring, plaster walls, exposed wooden structural elements and stone trim. (Architect: Paula Baker. Contractor: Prull & Associates, Inc. Photo: Lisl Dennis.)

Paints and Coatings

Paints and sealers can be a source of indoor air pollution. Certain specialty paints, on the other hand, can improve indoor air quality by sealing off undesirable substances so that they do not offgas into living space. Oil based paints have a much higher VOC output than water based latex paints. They have traditionally been considered to be more enduring than latex paints. However, given recent advances in the formulation of latex based paints, oil based paints can be avoided entirely in residential construction. Most latex based paints, although preferable to oil based paints, still have a high VOC content and contain harmful biocides and preservatives, such as formaldehyde, which extend their shelf life. Plant based paints are also available which contain natural ingredients rather than petrochemical products. However, some sensitive people find the citrus ingredients and other aromatics used in these paints unacceptable.

With increasing market demand for healthier paints, some major paint companies have recently developed very low VOC paints which are comparable in price to high quality standard latex paint lines. Appearing below are sources of healthier paints.

Paints and Coatings

Commercially Available Sources for Low VOC Paints

- **Coronado Supreme Collection**: VOC free latex paints.
- **Healthspec**: Low odor, durable, acrylic latex paint by Sherwin Williams.
- **Spred 2000**: Zero VOC, acrylic latex paint by Glidden Company.
- **Miller LBNF**: Specify low biocide, no fungicide paint.
- **Pristine**: Zero VOC, acrylic latex paint by Benjamin Moore & Company.
- **Spectra-Tone Paint**: Low VOC, no solvent, latex paints.

Specialty Hypo-allergenic Paints

- **AFM Safecoat Enamel** and **AFM Safecoat Zero VOC Paint**.
- **Ecology Paints**: Odorless, formaldehyde free, water based terpolymer paint.
- **Enviro Safe Paints**: Contain no fungicides and are low biocide, custom mixed to order.
- **Murco GF1000** and **Murco LE1000**: Contain no fungicides and only "in-can" preservatives which enhance shelf life, and then become entombed in the dry paint.

Primer Paints that Seal In and Block VOCs

- **AFM Safecoat Primer Undercoater**: For use on drywall, wood, and masonite. Seals and reduces outgassing.
- **BIN Primer Sealer**: A white pigmented shellac sealer used as an undercoat/primer/sealer. It is free of biocides, and will effectively seal in odors from drywall. It should be used in a well-ventilated space since the alcohol base is strong smelling during application. It is available through most paint and hardware stores.

Stains and Transparent Finishes

Most standard sealers for wood are solvent based and contain several highly toxic chemicals that outgas for long periods of time after application. Recently, several more healthful water based products have come on the market. Since water based products tend to raise the grain on wood or absorb unevenly, many installers who are inexperienced with water based products have been disappointed with the results. We have found several good installers who have overcome their initial reluctance and now insist on using less toxic, water based products, knowing that in doing so they are safeguarding themselves, their employees, and their clients. Natural, more healthful oils, lacquers, shellacs, and waxes are also available.

Appearing below are low or nontoxic finishes for wood that may be specified in the healthy home.

Stains and Transparent Finishes

Clear Seal Water Reducible Wood Finishes

- **AFM Safecoat Hard Seal** used in conjunction with **AFM Safecoat Lock-In New Wood Sanding Sealer**.
- **Aqua-Zar**: Water based nonyellowing polyurethane in satin or gloss finish.
- **Hydroshield Plus**: Water based polyurethane for interior and exterior use.
- **Pace Crystal Shield**: Clear sealant for flooring and woodwork; odorous when wet, odor free once dry.
- **Skanvahr** and **Kabinet**: Durable floor and cabinetry finish with low volatile organic compounds (VOCs).
- **Zip Guard Environmental Water Base Urethane**: Clear finish for interior woodwork.

Natural Oil, Lacquer and Shellac Wood Finishes

- **Auro No. 131 Natural Resin Oil Glaze**: Oil varnish, transparent tintable varnish. For exterior use.
- **Auro No. 213 Clear Semi-Gloss Shellac**: Semi-gloss, interior wood lacquer.
- **Auro No. 235 Natural Resin Oil Top Coat**: Oil lacquer, white, semi-gloss lacquer for interior and exterior use.
- **Auro No. 240 Natural Resin Oil Top Coat**: Interior and exterior in eight colors.

Wood Stains

- **AFM Safecoat Durostain**: Interior and exterior.
- **Bio Shield Earth Pigments #88**.
- **Hydrocote Danish Oil Finish**: Stain that colors and protects in one step.
- **Livos Kaldet Stain, Resin & Oil Finish**: Satin, semi-flat, water resistant finish, interior and exterior, strong surface hardening capacity for wood cabinets, doors, and windows.
- **Old Growth**: Two-step process using minerals and hydrogen peroxide. Gives an aged patina to wood.
- **OS/Color One Coat Only**: Natural oil based stains.

Clear Vapor Barrier Sealants for Wood

These products are used to help lock in noxious fumes so that they do not escape into the air. In fact, since no seal is ever perfect, vapor barrier sealants generally serve to decrease the amount of outgassing at any one time while increasing the overall time it takes for any substance to completely volatilize. We recommend that all efforts be made to speed up the outgassing time prior to application of vapor barrier sealants.

Outgassing can be accelerated by airing outdoors where protected from the weather or by using filtration, "bake-outs," or "adsorbers" indoors. VOCs readily release noxious vapors in heat. Harmful chemicals can thus be dissipated more quickly if they are exposed to elevated temperatures. VOCs can be baked out by repeatedly heating a space and then airing it out with fresh air. Adsorbers are substances such as zeolite or aluminum silicate to which VOCs adhere. When adsorbers are placed in a room, they help remove VOCs from the ambient air.

Although most coatings seal to some degree, the following products are advertised by respective manufacturers as recommended specifically because they lock in noxious fumes.

Clear Vapor Barrier Sealants for Wood

- **AFM Safecoat Hard Seal**: Clear sealer for low moisture areas.
- **AFM Safecoat Safe Seal**: Clear sealer for porous surfaces; also an effective primer.
- **Pace Crystal Shield**: Replaces lacquers, varathanes, and urethanes.

Manufacturer's instructions for application must be followed in order to achieve an optimum seal.

Endnotes

1. Thrasher, Jack and Alan Broughton, *The Poisoning of Our Homes and Workplaces* (Seadora, Inc., 1989), 50-72.
2. Bill Hirzy, EPA Senior Scientist, President of EPA Union Local 2050, "Chronology: EPA and Its Professionals, Union Involvement with Carpet," 1992. *Cited in* "Carpet: Trouble Underfoot," *Informed Consent* (November/December 1993), 31.
3. Environmental Protection Agency, "Indoor Air Quality and New Carpet: What You Should Know." (Washington, DC: U.S. Government Printing Office, EPA/560/2-91/003, March 1992.) Pamphlet.
4. Hirzy, p. 31.
5. Memorandum by Susan E. Womble, Project Manager, Consumer Products Safety Commission (CPSC) Chemical Hazards Program, "Evaluation of Complaints Associated with the Installation of New Carpet," August 13, 1990.
6. Ibid.
7. "Carpet Industry Program Steps Out Front on Indoor Air Quality: Labeling for Consumers Now Underway," Carpet and Rug Institute (Dalton, GA) press release, July 17, 1992.
8. "Carpet Off-gassing and Lethal Effects on Mice," Anderson Laboratories (Hartford, VT) press release, August 18, 1992.

Chart 9-1: Resource List

Product	Description	Manufacturer/Distributor
86001 Seal	Clear, water reducible sealer and primer for gypsum board.	Palmer Industries, Inc.; 10611 Old Annapolis Rd.; Frederick, MD 21701; (888)685-7244, (301)898-7848
AFM SafeChoice Carpet Guard	Sealer designed to help prevent outgassing of harmful chemicals from carpet backing and adhesives.	Same

Chart 9-1: Resource List

Product	Description	Manufacturer/Distributor
AFM SafeChoice Carpet Lock-Out	A final spray application that seals harmful chemicals in carpet and repels dirt and stains.	Same
AFM SafeChoice Carpet Shampoo	Odorless carpet shampoo that helps remove chemicals such as pesticides and formaldehyde from new carpet.	AFM (American Formulating and Manufacturing); 350 West Ash Street, Suite 700; San Diego, CA 92101-3404; (800)239-0321, (619)239-0321
AFM Safecoat Acrylacq	High-gloss clear water based wood finishes replacing conventional lacquer.	Same
AFM Safecoat Almighty Adhesive	Zero VOC clamping adhesive for gluing wood and wood laminates.	Same
AFM Safecoat DuroStain	Seven different earth pigment, semi transparent, wood stains. Interior/exterior, water based.	Same
AFM Safecoat Enamel	Comes in flat, eggshell, semi-gloss and gloss, water based paints without extenders, drying agents or formaldehyde.	Same
AFM Safecoat Hard Seal	Water based, general purpose clear sealer for vinyl, porous tile, concrete, plastics, and particle board plywood. Not recommended where exposed to heavy moisture or standing water.	Same
AFM Safecoat Lock-In New Wood Sanding Sealer	Sandable sealer helps raise grain on new woods in preparation for sanding prior to finish coat.	Same
AFM Safecoat MexeSeal	Topcoat used over Paver Seal .003 undercoat. Very durable sealer providing water and oil repellancy for use on Mexican clay tile, stone, granite, concrete, and stone pavers. Glossy when multiple coats are applied.	Same
AFM Safecoat New Wallboard Primecoat HPV	Water reducible, one coat coverage primer for new gypboard greenboard and high recycled content material.	Same
AFM Safecoat Paver Seal .003	Undersurface sealer for new or unsealed porous tile, concrete, and grout. Used with MexeSeal topcoat.	Same
AFM Safecoat Polyureseal	Clear gloss wood finish replaces conventional solvent and water based polyurethanes for low traffic interior wood floor and furniture applications.	Same
AFM Safecoat Polyureseal BP	As above for high traffic situations, high durability and abrasion resistance.	Same
AFM Safecoat Primer Undercoater	Primer for use on drywall, wood and masonite with excellent sealing properties; reduces outgassing.	Same

Chart 9-1: Resource List

Product	Description	Manufacturer/Distributor
AFM Safecoat Safe Seal	Clear sealer for porous surfaces effective in blocking outgassing from processed woods. Improves adhesion of finish coats.	Same
AFM Safecoat 3 in 1 Adhesive	Adhesive for ceramic, vinyl, parquet, Formica, slate, and carpet.	Same
AFM Safecoat Zero VOC Paint	Flat and semi-gloss paint. No VOCs, formaldehyde, ammonia, or masking agents.	Same
Aqua-Zar	Water based, nonyellowing polyurethane in satin or semi-gloss finishes.	United Gilsonite Laboratories; P.O. Box 70; Scranton, PA 18501-0070; (800)845-5227, (717)344-1202
Auro No. 131 Natural Resin Oil Glaze	Transparent, tintable finish for exterior porous wood protection. Note: This and the following Auro products are made with natural plant and mineral derivatives in a process called plant chemistry. Some sensitive individuals may have severe reactions to the natural turpenes, citrus derivatives, and oils.	Sinan Company; P.O. Box 857; Davis, CA 95617; (916)753-3104
Auro No. 211 - 215 Shellacs	Various natural shellacs with different degrees of gloss.	Same
Auro No. 235 & 240 Natural Resin Oil Top Coat	Indoor/outdoor enamels in white and eight colors.	Same
Auro No. 383 Natural Linoleum Glue	Adhesive for linoleum flooring.	Same
Auro No. 385 Natural Carpet Glue	Organic binder which remains permanently elastic.	Same
Bangor Cork Company	Natural cork "carpeting" and battleship linoleum flooring.	Bangor Cork Company; William and D Streets; Pen Argyl, PA 18072-1025; (610)863-9041
BIN Primer Sealer	White paint on vapor barrier sealer for use as prime coat on gypboard and wherever an opaque sealer is desired.	Wm. Zinsser & Company; 39 Belmont Drive; Sommerset, NJ 08875; (732)469-8100
Bio Shield Cork Adhesive	Water based elastic glue for cork, linoleum, or wool with jute backing. Adheres to concrete, wood, or plywood. This and the following Bio Shield products comprise a line of oil finishes free of formaldehyde, lead and other heavy metals, and fungicides. As with many natural products, some chemically sensitive individuals may not tolerate turpenes, oils, or citrus based and other aromatic components found in these formulations.	Eco Design/Natural Choice; 1365 Rufina Circle; Santa Fe, NM 87505; (800)621-2591, (505)438-3448

Chart 9-1: Resource List

Product	Description	Manufacturer/Distributor
Bio Shield Earth Pigments #88	Fine pigment powders extracted from earth or rock containing little or no heavy metals. Can be used with Bio Shield Oil Finishes.	Same
Bio Shield Hard Oil #9	For use on hardwood and softwood floors and stone in areas exposed to moisture. Use over surfaces primed with Bio Shield Hardwood Penetrating Oil Sealer #8 or Bio Shield Hardwood Penetrating Oil Sealer #5.	Same
Bio Shield Natural Resin Floor Finish #92	Finish for hardwood and softwood floors primed with Bio Shield Hardwood Penetrating Oil Sealer #8 or Bio Shield Penetrating Oil Sealer #5.	Same
Bio Shield Penetrating Oil Sealer #5	A sealer, undercoat, and primer for absorbent surfaces of wood, cork, stone, slate, or brick.	Same
Bio Shield Penetrating Oil Sealer #8	A thinner oil for priming less absorbent woods.	Same
C-Cure AR Grout	Portland cement, lime, earth pigment, and sand, without latex modifiers. Available in 40 colors.	C-Cure Corporation; 16225 Park Ten Pl., Suite 850; Houston, TX 77084-5155; (800)895-2874, (713)492-5100
C-Cure Floor Mix 900	Dry-set mortar used for floor and wall installations of absorptive, semi-vitreous and vitreous tiles.	Same
C-Cure Multi-Cure 905	Latex-cement mortar used for setting all types of ceramic tile; used on dry interior walls and exterior grade plywood.	Same
C-Cure PermaBond 902	For low odor tile setting.	Same
C-Cure Supreme 925 Grout	Dry tile grout with exceptional working qualities and a permanent joint life; nonshrinking, nontoxic, odorless, inhibits fungal growth.	Same
C-Cure ThinSet 911	Dry-set mortar used for installation of low absorptive tiles (less than 7%).	Same
C-Cure Wall Mix 901	Dry-set mortar used for the installation of absorptive tiles (more than 7%).	Same
Cemroc	A lightweight, strong, noncombustible, highly water resistant board that can be used as a backer for ceramic tile installations in wet areas in place of greenboard.	Eternit, Inc.; Village Center Drive; Reading, PA 19607; (800)233-3155, (215)926-0100
Ceramic Tile Grouts	Grouts contain Portland cement and silica sand. Available in 35 colors in sanded and unsanded.	Nontoxic Environments; P.O. Box 384; Newmarket, NH 03857; (800)789-4348, (603)659-5919

Chart 9-1: Resource List

Product	Description	Manufacturer/Distributor
Coronado Supreme Collection	Zero VOC latex paint; dries to highly washable surface.	Coronado Paint Co.; P.O. Box 308; Edgewater, FL 32132-0308; (800)883-4193, (904)428-6461
DLW Linoleum	Manufactured from all natural products (linseed oil, cork, wood flour, resin binders, gum, and pigments) with natural jute backing.	Gerbert Ltd.; 715 Fountain Ave.; Lancaster, PA 17601; (800)828-9461
Eco Design/Natural Choice	Source for cork floors and other natural building products.	1365 Rufina Circle; Santa Fe, NM 87505; (800)621-2591, (505)438-3448
Ecology Paints	Water based, resin terpolymer paint; odorless, formaldehyde free. Also carries Canary line of paints which are biocide and fungicide free.	Innovative Formulations; 1810 S. 6th Avenue; Tucson, AZ 85713; (800)346-7265, (520)628-1553
Endurance II	Synthetic jute pad; odorless, hypoallergenic.	Distributed through Statements; 1441 Paseo de Peralta; Santa Fe, NM 87501; (505)988-4440
Enviro Safe Paints	No fungicide, low biocide paints mixed to order.	Chem Safe; P.O. Box 33023; San Antonio, TX 78265; (210)657-5321
Envirobond #801	Water based organic mastic; VOC compliant.	WF Taylor Company Inc.; 11545 Pacific Avenue; Fontana, CA 92337; (800)397-4583, (909)360-6677
Envirotec Health Guard Adhesives and Seaming Tapes	A line of zero VOC, solvent free adhesives and seaming tapes without alcohol, glycol, ammonia, or carcinogens. Call distributor to find best product for a particular installation.	Same
Forbo Industries	Natural linoleum and cork flooring.	Forbo Industries; P.O. Box 667; Hazelton, PA 18201; (800)233-0475, (717)459-0771
Hardibacker Board	Cementitious tile backer board for use in moist/wet applications.	James Hardie Company; 10901 Elm Avenue; Fontana, CA 92335; (800)426-4051, (714)356-6300
Hartex Carpet Cushion	Odorless, synthetic jute underpadding for carpet.	Leggett & Platt; 1100 S. McKinney Street; Mexia, TX 76667; (800)880-6092, (817)562-2814
HealthSpec	Low odor interior latex paint. More durable but a bit more odorous than certain other commercially available low/no odor paints.	The Sherwin Williams Company; 101 Prospect Avenue NW; Cleveland, OH 44115; (800)321-8194, (216)566-2902. Sold at Sherwin Williams paint stores throughout the country.
Hendricksen Naturlich	Manufacturer of wool carpeting, other natural fiber carpeting, felt underpads, and adhesives.	Hendricksen Naturlich; P.O. Box 1677; Sebastopol, CA 95473; (707)824-0914; Fax (800)329-9398
Hydrocote Danish Oil Finish	Nontoxic penetrating oil. One-step stain and seal in nine wood tones.	The Hydrocote Company Inc.; P.O. Box 160; Tennent, NJ 07763; (800)229-4937, (908)257-4344

Chart 9-1: Resource List

Product	Description	Manufacturer/Distributor
Hydroshield Plus	Clear coat available in gloss or satin sheen. Water based polyurethane giving impact and weather resistance for interior or exterior.	Same
Kabinet	Extremely low odor, clear finish for cabinetry.	Skanvahr Coatings Ltd.; 8311 Market; Spokane, WA 99207; (800)329-3405, (509)466-1841
Laticrete Additive Free Thinset	For use over concrete, cement backer board, or wire reinforced mud. Contains Portland cement and sand.	Nontoxic Environments, Inc.; P.O. Box 384; Newmarket, NH 03857; (800)789-4348, (603)659-5919
Living Source	Source of nontoxic carpets and adhesives.	For consultation, contact Living Source; P.O. Box 20155; Waco, TX 76702; (800)662-8787, (817)776-4878
Livos Ardvos Wood Oil	Penetrating oil primer and finish for interior hardwoods. May be topcoated with Livos Bilo Floor Wax. This and the following four items comprise a line of plant chemistry products made from plant and mineral derivatives. As with many natural products, some chemically sensitive individuals may not tolerate turpenes, oils, citrus based and other aromatic components found in these formulations.	Eco Design/Natural Choice; 1365 Rufina Circle; Santa Fe, NM 87505; (800)621-2591, (505)438-3448
Livos Bilo Floor Wax	A clear, mar resistant finish for wood, stone, terra cotta, and linoleum.	Same
Livos Glievo Liquid Wax	Clean, apply, and buff furniture and floor wax. We have also applied this product to plastered walls.	Same
Livos Kaldet Stain	A stain and finish oil in 12 colors for interior and exterior surfaces made of wood, clay, or stone.	Same
Livos Meldos Hard Oil	A penetrating oil sealer and finish for interior absorbent surfaces made of wood, cork, porous stone, terra cotta tiles, and brick.	Same
Mapei 2-1/2 to 1	An additive free grout for joints larger than 3/8".	Mapei Inc.; 1501 Wall Street; Garland, TX 75041; (800)621-6491, (214)271-9500
Medex	Formaldehyde free, exterior grade, medium density fiberboard.	Medite Corporation; P.O. Box 4040; Medford, OR 97501; (800)676-3339, (503)773-2522
Medite 11	Formaldehyde free, interior grade, medium density fiberboard.	Same
Miller LBNF	LBNF line which contains low biocide content and no fungicides. Solvent free. Flat, satin, and semi-gloss.	Miller Paint Company; 317 SE Grand Avenue; Portland, OR 97214; (800)852-3254

Chart 9-1: Resource List

Product	Description	Manufacturer/Distributor
Murco GF1000	Flat wall paint. Odorless when dry. Preservatives are entombed in dry paint. No slow releasing compounds or airborne fungicides.	Murco Wall Products; 2032 N. Commerce; Fort Worth, TX; (800)446-7124, (817)626-1987
Murco LE1000	Higher gloss paint formulated as above for use where latex enamels are recommended.	Same
Murco M-100	Powdered all purpose joint cement, a texture compound formulated with inert fillers and natural binders only. No preservatives.	Same
Natural Cork C°	Natural cork flooring in a variety of colors and finishes.	Natural Cork C Ltd.; 1750 Peachtree St., Ste. 305; Atlanta, GA 30309; (800)404-2675, (404)872-4168
Naturel Cleaner/Sealer	Nontoxic, biodegradable, water soluble flakes that clean, protect, and finish stone surfaces.	Available by mail order from Solutions; P.O. Box 6878; Portland, OR 97228; (800)342-9988
Okon Seal & Finish	Satin or gloss. Clear sealer that can be used to seal plaster.	Okon Inc.; 6000 W. 13th Avenue; Lakewood, CO 80214; (800)237-0565, (303)232-3571
Old Growth, Aging and Staining Solutions for Wood	Wood is treated with a nontoxic mineral compound and then with a nontoxic catalyst that binds the natural mineral colors to cellulose creating an aged patina. It imparts antimicrobial and antifungal properties to the wood while the pigments provide UV protection.	CrossLink; P.O. Box 1371; Santa Fe, NM 87504-1371; (888)301-9663, (505)983-6877
OS/Color One Coat Only	Twelve different stain colors in base of vegetable oils. Interior/exterior use. No preservatives or biocides.	Ostermann & Scheiwe; P.O. Box 669; Spanaway, WA 98387; (800)344-9663
OS/Color Hard Wax/Oil	A satin matt oil/wax finish for interior wood floor and cork. Water repellant, easy to refinish.	Same
Pace Crystal Shield	Replaces lacquers, varathanes, and urethanes. Clear seal strong enough for hardwood floors. Can be used as sealant to block formaldehyde and other chemical emissions from manufactured wood products. Can be used to seal tile flooring.	Pace Chem Industries; P.O. Box 1946; Santa Ynez, CA 93460; (800)350-2912, (805)686-0745
Pristine	Commercially available acrylic latex paint without VOCs. Available in several finishes.	Benjamin Moore & Company; 51 Chestnut Ridge Road; Montvale, NJ 07645; (800)344-0400, (201)573-9600

Chart 9-1: Resource List

Product	Description	Manufacturer/Distributor
Skanvahr	Extremely low odor, waterborne, clear, nonyellowing, self-cross linking wood finishes for flooring. Products for high traffic and residential use with satin, semi-gloss, and gloss.	Skanvahr Coatings Ltd.; 8311 Market; Spokane, WA 99207; (800)329-3405, (509)466-1841
Sodium Silicate	Clear sealer for concrete floors. Widely distributed in hardware stores.	Ashland Chemical Inc.; 5200 Blazer Pkwy.; Dublin, OH 43017; (800)258-0711, (614)889-3333
Spectra-Tone Paints	Zero VOC, solvent free, latex paint; cost competent with standard quality latex.	Spectra-Tone Paint Corporation; 1595 San Bernardino, CA 92408-2946; (800)272-4687, (909)478-3485
Spred 2000	Commercially available paint without petroleum based solvents. Zero VOCs.	Glidden Company; 925 Euclid Avenue; Cleveland, OH 44115; (800)221-4100, (216)892-2900
Tu Tuf 3	High density, cross laminated polyethelene, puncture resistant air barrier.	Stocote Products Inc.; Drawer 310; Richmond, IL 60071; (800)435-2621, (815)675-6713
Zip Guard Environmental Water Base Urethane	Clear finish for interior woodwork.	Star Bronztec; P.O. Box 2206; Alliance, OH 44601; (800)321-9870, (330)823-1550

Further Reading and Services

Anderson Labs, P.O. Box 323, West Hartford, VT 05064, (802)295-7344. For evaluation of toxic effects of selected carpets, insulation, and other building materials through testing on mice. Consultations are available by phone for a fee.

"Carpet and Indoor Air: What You Should Know." June 1993. Available free of charge from New York State Attorney General, 120 Broadway, New York, NY, 10271.

Environmental Access Research Network (EARN), 315 W. 7th Avenue, Sisserton, SD 59645. For a list of carpet related articles, studies, and reports available from EARN's photocopying service, send $1.00 and request "Carpet List."

Thrasher, Jack and Alan Broughton. *The Poisoning of Our Homes and Workplaces, The Truth about the Indoor Formaldehyde Crisis.* Seadora, Inc., 1989.

Division 10 - Specialties

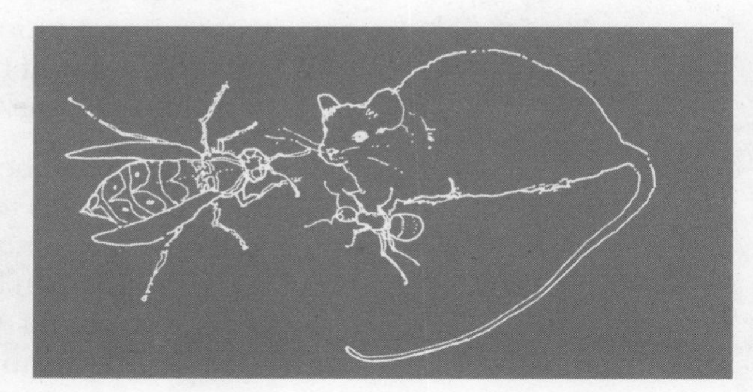

Integrated Pest Management

Integrated Pest Management and New Home Construction

All creatures have their rightful place in nature. However, this place is not within the walls of human habitations. Hence the need for humans to exercise pest control. While many pest control companies advocate regular prophylactic spraying of homes with toxic chemicals, this approach can have devastating consequences to the health of all living beings, including the occupants of the home. Although insects are effectively killed, the underlying structural problems that created inviting conditions for pests have not been addressed, so the pests eventually return.

Integrated pest management (IPM) offers a holistic approach to controlling pests. IPM differs from standard pest management in that the emphasis is on prevention and the least toxic methods of pest control. The goal is to work effectively with nature to alter conditions without creating harm to the environment. IPM precepts are summarized below.

- Accurate identification or "naming" of a pest is necessary so that its modus operandi may be understood and incorporated into a pest management plan.
- Careful consideration is given as to whether any action at all is required. "Entomophobia" is rampant in our culture. For many, the first reaction upon seeing an insect is to kill it. Pesticide commercials persuade us that panic and lightning speed action are necessary. In

contrast, IPM encourages an attitude of tolerance to creatures that do no harm. It also encourages rational determination as to when intervention will be necessary.

- If a pest must be eliminated, the first step is to see if its current access to nourishment and habitat can be limited. In the case of ants, for example, this might mean cleaning up crumbs from the floor and counters and caulking the cracks.
- If a pest must be trapped or killed, then the most environmentally benign methods are considered first. Least toxic chemicals are employed as a last resort.
- If a chemical must be used, then toxicity, risk, and exposure must be carefully evaluated. (Refer to Northwest Coalition for Alternatives to Pesticides-NACP for fact sheets on the various pesticides.)
- Careful observation and record keeping are an essential part of an integrated pest management program.

In new home construction you have the opportunity and responsibility to prevent infestations before they occur. An integrated approach to pest management in new construction would include the following items.

- Identification of potential pests found in the building site area.
- Research on identified pests, including eating habits, reproductive cycles, habitat, and common routes of entry into the home.
- Use of strategies in home construction that will create inhospitable and inaccessible conditions for pests.

In general, a well-constructed home will also be pest resistant, incorporating the following features:

- Weather tightness.
- Appropriate grading and drainage.
- Provisions made for the prevention of excess moisture buildup from within. These provisions include extraction fans and windows that allow cross ventilation.
- Dry wood without rot or infestation used in construction.
- Exterior wood appropriately treated for prevailing climatic conditions.
- All openings screened.
- Ground cover, leaves, chip and wood piles, and other potential insect habitats will be kept at a distance from the building.

Throughout the book we have specified techniques for the prevention of pests where appropriate. If you are building in an area characterized by a particularly difficult pest problem, then you may need to take measures above and beyond the scope of this book. For example, if your home is next to a shipyard or close to a row of poorly constructed grain elevators, then you may wish to incorporate more rat control techniques into your construction than would generally be specified. We heartily recommend *Common Sense Pest Control* by Olkowski, et al. as a comprehensive guide to specific pest problems. The following chart provides an overview of major household pests and construction techniques that discourage them.

Chart 10-1: Common Pests and Management Strategies

Pest	Types of damage	Modus operandi	Recommendations
Termites (subterranean)	• Structural damage • Tunnels created in wood	• Require moist conditions • Must be able to get from the soil into wood structure via earthen tubes; these insects do not live in wood	• Control moisture • Seal off wood from ground contact • Use termite shielding, and/or termite resistant sill plates
Termites (drywood)	• Structural damage • Tunnels created in wood	• Can access house through walls • Live in wood	• Tight construction • Caulked joints • Boric acid in framing
Rats	• Unaesthetic • Carry disease • Destroy food supply • Breed quickly	• Require hole 1/2" wide to enter	• Screen all points of entry including openings along pipes and wires • Make home weather tight • Ground floors should be elevated 18" above ground • Subterranean concrete floors should be a minimum thickness of 2" • Use wire mesh under wood floors • Use noncombustible cement stops between floor joists
Mice	• Chew through electrical wires causing fire hazard • Transmit pathogens • Breed quickly	• Require openings the size of a dime • Feed on dry foods, grains, clothing, paper • Usually inhabit buildings when outdoor climatic conditions become severe	• Seal all holes and crevices, especially where pipes and wires protrude through surfaces
Ants (carpenter)	• Create nests inside walls, ceilings, under siding, and where wood and soil are in contact near foundations • Infest both hardwood and softwood	• Require high moisture content wood (minimum 15%)	•Use kiln or air dried lumber and keep it dry •Prevent structural wood and earth from coming into contact with each other •Allow for proper ventilation of damp areas
Bees (carpenter)	• Chew on wood • Burrow into structural members and exposed wood elements	• Enjoy untreated exposed wood (especially softwoods)	• Paint or varnish exposed wood (sills, trim, etc.) • Fill in holes and indentations in wood

Chart 10-1: Common Pests and Management Strategies

Pest	Types of damage	Modus operandi	Recommendations
Beetles (wood-boring)	• Bore through wood	• Require moisture content in wood to be 10 to 20%	• Prevent moisture changes and temperature fluctuations • Allow for good ventilation in attic spaces • Keep roof frame and sheathing dry • Use air or kiln dried lumber • Seal wood
Cockroaches	• Invade food storage areas such as kitchens and cupboards • Can carry disease causing organisms	• Most species prefer warm, moist areas	• Avoid moisture and decayed organic buildup in or near home • Use boric acid in framing in areas prone to infestation • Use screens on vents and windows
Fungi (wood decay)	• Attacks and weakens wood leaving it susceptible to invasion by wood boring and eating insects	• Grows best at temperatures between 50 and 95 degrees F • Requires a minimum of 20% moisture	• Allow for proper roof insulation and ventilation to prevent condensation • Seal wood joints at corners, edges, and intersections • Prevent moisture accumulation near pipes, vents, and ducts • Do not use wood in containing mold or mildew in construction • Seal all wood exposed to the elements • Control moisture buildup generated by human activity through proper ventilation strategies • Use building products and procedures that allow moisture vapor to escape rather than being trapped

Further Reading and Resources

Books

Moses, Marion. *Designer Poisons: How to Protect Your Health and Home from Toxic Pesticides.* San Francisco, CA: Pesticide Education Center, 1995. A sobering expose about specific pesticides and the chronic health effects that can result from their use; provides useful information on safer alternatives.

Olkowski, William, Sheila Daar, and Helga Olkowski. *Common Sense Pest Control: Least-toxic Solutions for Your Home, Garden, Pets, and Community.* Newtown, CT: Taunton Press, 1991. Comprehensive, well-documented information on integrated pest management and least-toxic pest control for all kinds of pests.

Schultz, Warren. *The Chemical-Free Lawn: The Newest Varieties and Techniques to Grow Lush, Hardy Grass.* Rodale Press, 1989. Techniques for growing lush and hardy grass without using pesticides, herbicides, or chemical fertilizers.

Organizations

Biointegral Resource Center (BIRC), P.O. Box 7414, Berkeley, CA 94707. Tel: (510)524-2567. Useful source of pesticide information.

National Coalition Against Misuse of Pesticides, 701 E. Street SE, Suite 200, Washington, DC 20003. Tel: (202)543-5450. E-mail: *ncamp@igc:apc.org.* Provides useful information about pesticides and nontoxic alternatives.

Northwest Coalition for Alternatives to Pesticides (NCAP), P.O. Box 1393, Eugene, OR 97440. Tel: (541) 344-5044. Provides a comprehensive information service on the hazards of pesticides and alternatives to their use. The NCAP maintains an extensive library of over 8,000 articles, government documents, videos, and other reference materials, and offers information packets, fact sheets, and the quarterly *Journal of Pesticide Reform.*

Product

Spotcheck Pesticide Testing Kit. Instant pesticide checking kit based on technology developed for U.S. military applications. Ships with supplies to carry out four tests of soil, water, food, or surfaces. Pesticides that the user can test for include Sevin, Dursban, Diazinon, Malathion, and Parathion. Available through The Cutting Edge Catalog; P.O. Box 5034; Southhampton, NY 11969; (800)497-9516.

Division 11 - Equipment

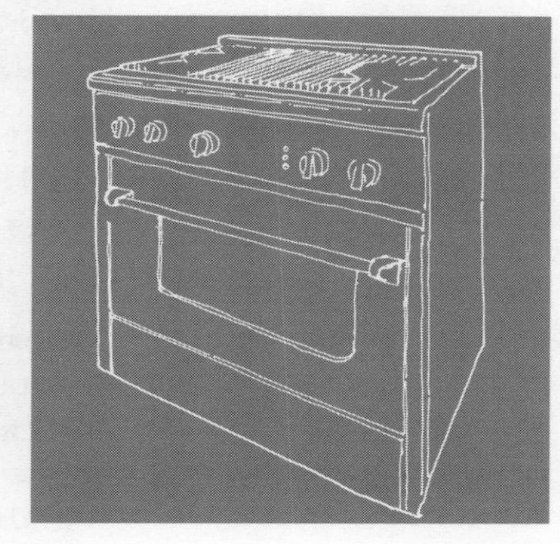

Water Treatment Equipment

Water Purification in Standard Construction

Poor indoor air quality is not the only form of pollution that affects human health. The water supply has also become increasingly polluted. Whether you are on a well or municipal water system, you may be receiving water that is unfit to drink. Water purification is not standard in home construction, and unless you specify water testing and purification, they will not be included. Although water purification is usually considered an "extra," whole house systems are best planned for and installed at the time of construction. Appearing below are some interesting facts that you should know about household water in the United States.

- Of the over 2,000 contaminants found in potable water, the U.S. Environmental Protection Agency (EPA) has established standards for only 83.[1]
- Municipal water is tested for an average of less than 30 contaminants.[2]
- Municipal water treatment is primarily set up for disinfection rather than water purification. Only 50 of the more than 60,000 water treatment facilities in the United States have modern equipment that can effectively remove toxic chemicals.[3]
- Over half of community water systems fail to comply with federal testing requirements.[4]
- Taking a 15-minute bath in water that contains chlorine or other VOCs allows as many toxins to enter an adult's body

as drinking two gallons of the same water. The absorption levels are even higher in children.[5]

- Forty-two million people in the United States are drinking water contaminated at dangerously high levels.[6]
- One in six people drinks water containing excessive amounts of lead which can impair cognitive function, especially in children.[7]
- Over the past two decades reports containing warnings about water pollution have been issued by the American Chemical Society, the National Academy of Science, the Natural Resources Defense Council, the American Petroleum Institute, and the Office of Technology Assessment.[8]

We strongly recommend a whole house water purification system as an essential feature of the healthy home. Choosing the proper system will depend on several factors including location, budget, water use, and taste preference. If you are receiving municipal water, laws have been passed to help determine what is in your water and how to best protect your household. The Safe Drinking Water Act of 1974, as amended in 1986, requires public water utilities to reveal to the public the following facts.

- Where your water comes from
- How the water is treated
- The nature of water quality tests, or what the water is tested for
- How people are notified when violations occur
- History of utility's water problems

You can obtain this information by contacting your local water utility company. Remember that the water leaving the treatment plant may be further contaminated by the time it reaches your tap. During its journey, treated water can pick up lead from solder and copper from metal pipes. Pipes made from PVC release chlorinated compounds and other chemicals into the water.

Municipally treated water is usually low in biological contaminants, but it is not well screened for industrial and hazardous waste. The chlorine with which almost all municipal water has been treated often reacts with naturally occurring organic compounds, creating potentially harmful organochlorides and trihalomethanes.

Testing your municipal water supply will render inconsistent results and may be unnecessary. Water experts in your area will be well informed on the range of contaminants found in your system and the best strategies for eliminating them.

In contrast, well water must be tested on a case by case basis. Two wells side by side can have very different water quality. Well water is free of added chlorine, but is consequently more susceptible to biological contamination. Testing for all possible contaminants is not financially feasible or necessary. Local experts will be able to advise you as to which tests are most appropriate in your area.

Chart 11-1: Common Contaminants Found in U.S. Water Supplies

Contaminant	Health effects (partial list)	Cause	Solution	Comments
Biological contaminants				
Bacteria and viruses	Flu-like symptoms, muscle aches, fatigue	Naturally occurring	Reverse osmosis (R/O) with TFC membrane, KDF, distillation, ultraviolet sterilization, ozonation, pasteurization	Not all solutions are equally effective for all biological contaminants. Consult a water treatment specialist.
Parasites	Gastrointestinal, diarrhea	Naturally occurring	Same	Same
Amoebae	Nausea, vomiting, dysentery	Naturally occurring	Same	Same
Molds		Naturally occurring	Same	Same
Algae		Naturally occurring	Same	Same
Organic contaminants				
Petroleum hydrocarbons	Cancer	Leaks from service stations, oil refineries, gasoline spills	Activated carbon block removes many organic compounds. KDF filters work well for chlorinated compounds.	Carbon filters must be changed regularly to be effective.
Benzene	Immune system sensitizer	By-product of gasoline use	Same	Same
Chlorobenzene	Cancer		Same	Same
Pesticides	Cancer, neurotoxicity	Agricultural runoff	Same	Same
Chlorinated compounds	Cancer, respiratory problems, chronic skin irritation	By-product of chlorination, released from PVC pipe	Same	Same
Trihaliomethanes	Cancer	By-products of chlorination	Same	Same

Chart 11-1: Common Contaminants Found in U.S. Water Supplies

Contaminant	Health effects (partial list)	Cause	Solution	Comments
Inorganic contaminants				
Nitrates	Break down to form carcinogens, interfere with ability of red blood cells to transfer oxygen	Nitrogen combines with soil bacteria to form naturally occurring nitrates. Nitrates from agricultural fertilizers leach into ground water.	Ion exchange, R/O, distillation	Particularly dangerous to infants. Has been linked to miscarriage.
Arsenic	Poison, carcinogen	Natural occurrence, pesticides, industrial pollutants	R/O, distillation, KDF	
Asbestos	Gastrointestinal cancer	Asbestos reinforced supply pipe	R/O, distillation, carbon block	Your utility company will know if asbestos reinforced supply pipe is used in your area.
Heavy metals (cadmium, chromium, selenium, mercury, lead, barium, aluminum, fluoride)	Health effects are specific to metal and include kidney damage, cancer, impairment of nervous system, behavioral disorders, lowering of IQ in children, Alzheimer's disease	Naturally occurring, plumbing piping and solder, automobile exhaust, industrial waste, fertilizers, pesticides	R/O, carbon block, distillation, and KDF are all used in specific circumstances	Carbon block is only effective when filter is relatively unclogged and water is continually tested. If concentrations are high enough there may be no filtration system that will produce safe, potable water.
Radon gas	Cancer, immune dysfunction	Radioactive gases permeate ground water and become airborne through household use.	Specially designed activated carbon	Whole house granular activated carbon filters can be designed to remove 90% of radon. With use, the filter will become contaminated and therefore should be located at a distance from living areas. Disposal of contaminated filter is a concern.
Radium and uranium	Bone cancer, leukemia, thyroid cancer	Caused by water dissolving radiologicals in soil or rock	Specialized R/O filters, ion exchanger	

Types of Water Purification Systems

No single filtration medium can remove all contaminants from all water. Because water quality and individual needs vary, no single combination of systems will provide a universal solution.

Choosing a system can be a complex and confusing process. The average homeowner typically does not know the right questions to ask in order to get accurate information. Water filtration systems have become popular network market items. Many people selling the filter devices are not much more knowledgeable about the range of needs and possibilities than potential clients. We recommend that you consult with an individual who possesses the following credentials.

- A broad based, longstanding experience with water quality in your area
- A wide variety of equipment from several manufacturers
- Ability to provide you with several options at various prices
- Ability to explain the pros and cons of each system

Chart 11-2: Summary of Water Filtration Methods

Type of system	How it works	What it eliminates	What it does not eliminate	Comments
Carbon filters. There are over 500 varieties of carbon filters. The two most commonly used filters in water purification are described here. Generally, the carbon filter works like a honeycomb with acres of surface area that absorb contaminants. These filters are not bacteriostatic and will become contaminated with use. Inexpensive sediment prefilters will extend life. Inexpensive chlorine tests can indicate when to change the filter if used with chlorinated water. These filters can become a source of pollution if not changed often enough. Locate the tank away from inhabited areas when used to filter radon.				
GAC (granulated activated carbon)	Carbon is steam treated so that the surface becomes pitted, thereby increasing surface area and absorption capacity.	Trihalomethanes, dissolved gases including chlorines, most pesticides, many chemical pollutants, radon gas	Heavy metals, sediment, fluoride, viruses and bacteriologicals, dissolved solids, particulates including radioactive particulate matter	Requires that water have sufficient contact time with the filter. Because GAC can breed bacteriologicals, it is most effective when used with treated municipal water.
Carbon block	Powdered carbon is glued together to form a matrix structure that absorbs contaminants.	As above for GAC, heavy metals, particulate matter	Fluoride, nitrates, viruses, and bacteriologicals	Considered more effective than GAC if water conditions are within certain parameters. Will only remove heavy metals for a limited time period. Periodic testing is essential. Not recommended for most heavy metal removal. Glue content is a concern. Whole house or point source available. Limited gallonage.

Chart 11-2: Summary of Water Filtration Methods

Type of system	How it works	What it eliminates	What it does not eliminate	Comments
Reverse Osmosis. Most often comes with sediment pre-filter and carbon post-filter. Plastic bodies can be problem for individuals with petrochemical sensitivities.				
City membrane CTA (cellulose triacetate)	Water is forced under pressure through a fine membrane which screens out dissolved solids.	Dissolved solids (80 to 90%), heavy metals, asbestos, radioactive particles, some bacteria	Dissolved gases, some biological contaminants, sediment	Most suitable for pretreated municipal water in which biological contaminants are already low. Filter requires chlorinated water supply to prevent bacteriological decay.
Well membrane TFC (thin film composite)	Same as above.	Same as above; biological contaminants	Dissolved gases, sediment	Cannot be used with chlorinated supply water unless prefiltered with carbon.
Sterilization				
Ultraviolet purification	Ultraviolet ray penetrates membrane of microbe and inactivates it.	Biological contaminants	Dissolved gases, sediment	Does not provide residual disinfection. Sediments, iron, manganese, or turbidity will make system ineffective.
Oxidation with ozone, hydrogen peroxide, chlorine, or injected air	Oxidation "burns" contaminants.	Clarifies, deodorizes, precipitates metals, oxidizes, eliminates bacteria, viruses and organic matter	VOCs, pesticides, chlorine; does not remove anything from water	Use of chlorine as oxidizing agent not recommended from ecological and health standpoints.
Others				
KDF (kinetic degradation fluxation)	Chemical transformation of contaminants as they pass through KDF which disrupts metabolic function of bacteria.	Controls bacterial growth; some heavy metals, chlorine, biological contaminants	Thrihalomethanes, bacteria	Very effective when used as pre-filter, followed by carbon filter and then reverse osmosis. Does not work well in all PH conditions. Suitable for water with very low bacteriological count. Is bacteriostatic but not a bacteriocide.
Shower head filters	Small filter/shower heads screw into existing plumbing.	Chlorine, trihalomethanes (only if filter contains carbon)	Radiologicals, pesticides, gasoline, bacteriologicals	Very inexpensive. Does not require plumber to install.

Chart 11-2: Summary of Water Filtration Methods

Type of system	How it works	What it eliminates	What it does not eliminate	Comments
Distillation	Water is turned to vapor, condensed, and then collected.	Dissolved solids, microorganisms, nitrates, heavy metals, sediment, radioactive particulate matter	VOCs, dissolved gases including chlorine	Effective when used with carbon post-filter. High maintenance, low production. Flat taste. Metal bodied distillers may add aluminum or other heavy metals to water. May leach necessary minerals from the body.
Sediment filters	Can be settlement tank where water is siphoned off top after particulate sinks. Can be a filter medium.	Particulate matter, sand, dirt	Only removes particles	Most often used as pre-filter for other systems. Back flushing models are self-cleaning.

Water Conditioning

Water conditioners are used to improve the aesthetic quality of water including, color, corrosiveness, clarity, and hardness. They use a process of ion exchange to eliminate aesthetically undesirable substances from the water, such as calcium and magnesium which precipitate on fixtures, laundry, hot water heaters, dishwashers, shower stalls, sinks, and skin. Water conditioners can also be effective in removing certain metals such as iron and low levels of manganese, both of which can cause stains. The conditioned water is often referred to as "soft water."

Water conditioners work by exchanging a calcium or magnesium ion with either sodium or potassium. Sodium has traditionally been the exchange regenerate. When sodium is used in a conditioning system, we strongly recommend using separate piping from the water supply entry point of the home to all potable water distribution points and hose bibs. The elevated levels of sodium which occur in water softened this way are not desirable for human or plant consumption. In fact, water softened with sodium contains levels high enough to be considered an environmental because it is harmful to ground water when the sodium eventually works its way back to the water table.

Potassium has more recently been introduced as a regenerate for water conditioning. It is a healthier and more ecologically sound choice. Potassium is essentially a refined potash, and when returned to the ground water, it can serve as a fertilizer for many plants.

Most people have a shortage of potassium in their diets. The small amount ingested daily from water conditioned with potassium is about the equivalent of half a banana and can actually be a positive addition to your diet.

NOTE: If your water is extremely hard, potassium levels might be too high to be safely ingested, necessitating a split distribution system as described above for systems using sodium as a regenerate. For those who have a medical condition affecting electrolyte balance, blood pressure, or kidney function, we suggest you consult a physician before you consider purchasing a water conditioning system.

Flow rate is affected by both the size and design of the water softener and must be appropriately specified on an individual basis. Water conditioning systems can also be designed to remove sediment, chlorine, odor from hydrogen sulfide, and elevated levels of iron.

Case Study:

Bath water found to be culprit in copper toxicity case

F.W. is a 63-year-old woman who was seen by Dr. Elliott for a chronic vaginal discharge which persisted for five years. She had been previously evaluated by several healthcare practitioners for this problem. Although her gynecologist was unable to find evidence of a yeast or bacterial infection, she was nevertheless placed on a variety of antibiotics which seemed to exacerbate the problem.

During the interview, it was discovered that the patient's symptoms seemed to improve when she traveled. She went on to disclose that a rash she had throughout her body also improved while she was traveling away from home. She concluded that her symptoms were probably related to stress, although there were no obvious new stressors in her life which could have accounted for this peculiar reaction. When questioned about events in her life that took place around the time of onset of her symptoms, the patient remembered that she had moved into a new home approximately five years ago. Dr. Elliott suspected that the source of the patient's problem may have been the bath water since water was the only substance in contact with her vagina.

A water sample was sent to a laboratory for analysis. The results showed extremely high copper levels. Upon further inquiry, it was discovered that many water samples from the same part of town were also showing high copper levels. Apparently, the carbon dioxide in the water created enough of an acidic environment to dissolve the copper in the water supply piping.

The patient decided to install a whole house water filtration system which could be customized to remove carbon dioxide in the household water. Within a few days after installation, her rash and vaginitis disappeared. Because of evidence of excess copper stored in her body, the patient underwent a program of heavy metal chelation and vitamin and mineral supplementation. She is currently doing well and is without complaints. In a follow-up visit, she revealed that the greenish ring that had been present on the bathroom fixtures had disappeared.

Discussion

While the need for filtering the household drinking water may be obvious, this case study illustrates that an unrecognized source of toxic exposure may be bath water. Because skin is a large surface area, it allows for significant absorption of substances into the body from bath water. We do not suggest that you avoid tub bathing, which can be both pleasurable and therapeutic. Instead, we recommend that your water be filtered at the point of entry into the house.

Filtration systems are most effective when they are customized to fit both the homeowner's personal needs as well as local water conditions. These conditions can vary greatly from one location to another. Whether or not you decide to install a whole house water filtration system, we recommend that you have your water tested periodically.

Residential Equipment

Much has been written about the energy efficiency of appliances. Appliances account for as much as 30% of household energy usage. Thus, choosing wisely can greatly reduce energy consumption. Because many sources of information are available on appliance energy values, in this book discussion is limited to health issues related to appliance selection. (See "Further Reading" section for sources on reducing appliance energy consumption.)

Appliances and Magnetic Fields

All motorized equipment found in homes will generate magnetic fields when in operation. Some epidemiological studies have linked exposure to these magnetic fields with increased incidence of cancer, Alzheimer's disease, and miscarriage. Magnetic fields from properly wired appliances drop off very quickly in an exponential relationship to the distance from them. These fields can be easily measured with a small hand-held instrument called a gaussmeter which allows the user to determine the safe distance from an appliance.

The U.S. government has not yet set reasonable standards for safe exposure levels, nor has it taken a strong position regarding health effects of magnetic fields. However, various government documents state that if you are concerned, you can practice "prudent avoidance" of these fields. Recommended safe exposure limits set by U.S. experts range from 0.5 milligauss to 1,000 milligauss(!). The Swedish National Energy Administration has recommended that children should not be subjected to magnetic field levels greater than 3 milligauss. We suggest that "prudent avoidance" translates to avoiding prolonged daytime exposure to fields greater than 1 milligauss.

"Division 16 - Electrical" includes recommended specifications and information for designing and building a home to help prevent magnetic fields transmitted by household wiring from surpassing 0.5 milligauss. The simple guidelines listed below may be followed to limit exposure to magnetic fields from appliances.

- Design your home so that major appliances are located at a safe distance from sitting and sleeping areas. In doing so, remember that magnetic fields travel with ease through walls made of common building materials and that areas located out of sight behind an appliance are also exposed. For example, placing a refrigerator back to back with a bed, even though separated by a wall, will expose the sleeping person continually to an unacceptably high magnetic field.

- Duration of exposure is a factor as well as strength. A lower level exposure for long periods of time may be more harmful than brief high level exposures. For this reason, pay particular attention to fields that may be generated around sleeping areas.

- We recommend that you buy and learn to use a gaussmeter. With this device, you can determine safe distances from all appliances. For more information on choosing and operating a gaussmeter, see "Division 16 - Electrical."

- Check your home and appliances regularly with the gaussmeter to determine whether field levels have increased. Elevated fields can sometimes indicate that an appliance has developed dangerous ground faults or that it is about to fail. Early detection of these fields will also decrease the risk of fire or electrocution.

Appliances and Electric Fields

Some research has indicated that electric fields affect melatonin levels. Melatonin is a hormone that plays an important part in the regulation of circadian rhythms. During the day melatonin levels are naturally low. Circulating melatonin is broken down during exposure to sunlight. At night (in darkness), melatonin levels in the body rise. Electric field exposure appears to suppress melatonin production. Consequently, sleeping in an electric field can disrupt circadian rhythms, thereby resulting in sleep disorders, anxiety, and depression. "Division 16 - Electrical" will demonstrate how to limit electric fields generated from the electrical wiring in and around homes.

Whereas magnetic fields exist only when appliances are being used, electric fields are present as long as the appliance is plugged in. Unfortunately, few appliances are manufactured in a manner that results in low electric fields. Rewiring appliances so that they operate with reduced electric fields is possible, but requires the services of an electrician familiar with electric field shielding. Electric fields from appliances are relatively easy to control by following the suggestions listed below.

- Keep appliances unplugged when they are not in use, especially in the bedroom. Not only will this eliminate the electric field, it will also reduce the risk of fire. Although this practice is much more common in Europe, the American Association of Home Appliances and Underwriters Laboratories has also issued a warning stating that small appliances should be unplugged as a fire prevention measure.
- Avoid using extension cords around beds or areas where your family spends a lot of time. They tend to emit high electric fields when they are plugged in.
- Use a battery operated or wind-up clock next to the bed.
- Wire your bedroom so that the circuitry can be conveniently shut off when you go to sleep, thus eliminating electric fields altogether. Refer to "Division 16 - Electrical for details."

Appliance Selection

Microwave Ovens

Microwave ovens are high EMF emitters. They are designed to heat food by creating microwave energy high enough to vibrate molecules in the food until heat is produced. When they are in use, magnetic fields extend up to 12 feet. The actual microwaves produced during operation are supposed to be contained in the oven by internal shielding, but leaks can occur. We do not recommend the use of a microwave oven. If you decide to use one nevertheless, the following suggestions will make using them safer.

- Maintain a distance of four to 12 feet from the microwave oven while it is in use. This is especially important for children who might enjoy watching the food as it is cooking.
- Have your appliance professionally checked for microwave leakage on an annual basis. You can check for yourself on a more frequent basis with a less precise do-it-yourself tester. Any detected leakage is unacceptable. Micro-wave leakage standards in the United States are much less stringent than in some parts of Europe. Unfortunately, differences in the power supply prevent the use of European microwave ovens in the United States.
- Do not use a microwave that appears to be malfunctioning. Signs of this would include sparks flying, funny noises, fires, or the unit turning on or cycling when the door is open. If any of these occur, evacuate the area immediately. Do not take time to try to unplug the unit. Instead, quickly shut off the circuit breaker to the microwave. If you do not know which one it is, shut them all off. Only then is it safe to return to the room to unplug the microwave oven.
- The shielding on a microwave is delicate. A very small amount of damage can cause a complete shielding failure. Even a paper towel stuck in the door is enough to cause the microwave shielding to fail.
- Do not microwave food in plastic containers. Chemicals from the plastic can leach into the food. Some of these chemicals are known to disrupt the endocrine system.

Trash Compactors

Trash compactors are now commonplace in new homes. They can be convenient, but they can also be difficult to clean. When choosing a trash compactor, examine it carefully to be sure you will be able to reach into it easily for cleaning. Verify that accidental liquid spills inside the unit will be contained and not run under or behind the unit. You may want to have the trash compactor installed in such a way that it can be easily removed for cleaning.

Some trash compactors come with a deodorizer chamber. With the exceptions of baking soda and zeolyte, most deodorizers contain phenols, formaldehyde, or paradichlorobenzene, all of which should be avoided.

Refrigerators and Freezers

There are many styles of refrigeration units available. The self-defrosting models have a drip pan located somewhere under the unit. Some units have drip pans located in the back or mounted internally where they are inaccessible. When purchasing a unit, make sure the drip pan is easily accessible from the front and has adequate clearance underneath for ease in cleaning. The pan should be cleaned monthly to prevent odors or the growth of microorganisms. It is also important to keep the cooling coils clean so that they do not become coated with dust. Not only will this improve your air quality, but the unit will not have to work as hard to stay cold, which, in turn, will save energy.

Cook Tops, Ovens and Ranges

All electric cook tops, ovens, and ranges produce elevated magnetic fields. Surprisingly, it is frequently the built-in electric clock that is the largest source of fields, regardless of whether the equipment is gas or electric. Use a gaussmeter to determine the distance of the field.

The act of cooking generates significant amounts of indoor air pollution through vapors and airborne particulate matter such as grease. In addition, food particles left on burners are incinerated and release combustion by-products.

Gas fueled appliances are a significant source of indoor air pollution, releasing carbon monoxide, carbon dioxide, nitrogen dioxide, nitrous oxides, and aldehydes into the air.

In his recent book, *Why Your House May Endanger Your Health,* Dr. Alfred Zamm describes how gas kitchen ranges have been the hidden culprit in many cases of "housewives' malaise." According to Zamm, "A gas oven operating at 350°F for one hour, because of the inevitable incomplete combustion, can cause kitchen air pollution, even with an exhaust fan in operation, comparable to a heavy Los Angeles smog. Without the fan, levels of carbon monoxide and nitrogen dioxide can zoom to three or more times that."

Although cooks usually prefer to cook with gas, we recommend electric cooking appliances. If you choose a gas range and oven, the following measures will help reduce the amount of pollution.

- Choose an appliance with electronic ignition instead of pilot lights. Any model built in the United States after 1991 will be equipped with electronic ignition.
- Have flames adjusted to burn correctly. They should burn blue. A yellow flame indicates incomplete combustion and the subsequent production of carbon monoxide.
- Follow the guidelines for proper ventilation discussed below.

Various smooth cook top surfaces are available including magnetic induction and halogen units. Because they are much easier to clean than coiled elements, they produce less pollution caused by the burning of trapped food particles. These units should be tested with a gauss meter to determine the extent of their magnetic fields while in operation.

Oven cleaning is another source of pollution generated in the kitchen. Continuous cleaning ovens contain wall coatings which continuously offgas noxious fumes. Self-cleaning ovens produce polynuclear aromatic hydrocarbons which are a source of air pollution. Most brand name oven cleaners are toxic. The safest way to clean an oven is with baking soda and elbow grease. If baking soda is poured over the spill shortly after it occurs, it can be easily cleaned up after the oven has cooled.

Kitchen Ventilation

Because the kitchen generates significant indoor pollution, the ventilation of this room should be given special consideration above and beyond general home ventilation. We recommend the largest range hood available with variable speed control so that you can adjust speed according to your requirements. Some models come equipped with remote fans which are quieter when operating.

Range hoods must be vented to the outside. There are models available that simply circulate the air through a carbon filter and back into the room. These do not sufficiently remove pollution created in the kitchen. Unfortunately, many kitchens come equipped with this type of unit because they are inexpensive and do not require a roof penetration.

Design the kitchen with an operable window or another source of air supply to offset the air lost through the operation of exhaust fans. If a clean source of air intake is not provided by design, the exhausted air will create negative pressure and air will then be sucked into the house through the path of least resistance. This path could be through a chimney flue for a fornace or water heater, causing dangerous backdrafting of air many times more polluted than that which it is replacing.

Laundry Appliances

Washers and dryers with porcelain on steel or stainless steel interiors are preferable to those with plastic interiors. Although gas dryers are more energy efficient than electric dryers, they cause the same pollution problems as gas ranges. By planning a laundry room with easy access to a drying yard, you can take advantage of the most energy efficient of all dryers, the sun.

Dryers should be vented directly to the outdoors. Some heat recovery devices are available which recirculate the hot air from the dryer back into the house. These do not sufficiently filter out fine particles and are not recommended.

Central Vacuum

Conventional portable vacuum cleaners suck air through a filter bag and then the "cleaned" air is pumped back into the room. The air that is returned is only as clean as the filtering mechanism allows. In fact, conventional vacuuming can stir up dust and pollen to such an extent that the ambient air is more polluted with small particulate matter than before the cleaning. High efficiency HEPA vacuums have recently become available. These filtering methods effectively trap microscopic particulate matter and are far superior to conventional vacuum cleaners.

Water filter vacuums were popular before the availability of HEPA vacuums. They can become a reservoir for mold and bacteria unless thoroughly dried after each use.

If you are building a new home, you have the opportunity to install a central vacuum system. When the motor and dirt receptacles are located remotely in a basement, garage, or utility room, central vacuums avoid the pollution problems associated with most portable models. Although more expensive than conventional portables, they cost only slightly more than a good HEPA or water filter model. They are convenient and easy to operate. The hose is simply plugged into a wall receptacle and there is no machinery to lug around. We recommend central vacuums that exhaust air directly to the outdoors.

Endnotes

1. Debra Lynn Dadd, *Nontoxic, Natural, and Earthwise* (Tarcher/Perigee, 1990), p. 41.
2. Ibid.
3. Ibid.
4. Ibid.
5. Quote by Greg Friedman, Good Water Company, Santa Fe, NM.
6. Maury M. Breecher and Shirley Lynde, *Healthy Homes in a Toxic World: Preventing, Identifying, and Eliminating Hidden Health Hazards in Your Home* (John Wiley & Sons, 1992), p. 121.
7. Ibid., p. 142.
8. Ibid.

Further Reading and Resources

Books

American Institute of Architecture, Denver Chapter. *Sustainable Design Resource Guide.* ADPSR Colorado, OEC Colorado, 1994. For information about energy efficiency and appliances.

Goldbeck, David. *The Smart Kitchen: How to Create a Comfortable, Safe, Energy-Efficient, and Environment Friendly Workplace.* Ceres Press, 1994.

Ingram, Colin. *The Drinking Water Book, A Complete Guide to Safe Drinking Water.* Ten Speed Press, 1991. A guide for safe drinking water.

Lono Kahuna Kapua A'o. *Don't Drink the Water, The Essential Guide to Our Contaminated Drinking Water and What You Can Do About It.* Kali Press, 1996.

Zamm, Alfred. *Why Your House May Endanger Your Health.* Simon and Schuster, 1980.

Resources

The Good Water Company, owned by Greg Friedman, 2778 Agua Fria, Bldg. C, Ste. B, Santa Fe, NM 87501. Tel: (800)471-9036, (505)471-9036. Water filtration and consultation.

Hague Quality Water International at 4343 South Hamilton Road, Groveport, OH 43125-9332. Tel: (614)836-2195. Excellent whole house water purification system.

Nigra Enterprises, owned by Jim Nigra, 5699 Kanan Road, Agoura, CA 91301-3328. Tel: (818)889-6877. Broker for high quality air and water filtration, vacuum cleaners, and heaters. Free consultation available.

Watercheck National Testing Laboratories, Inc. at 6555 Wilson Mills Road, Cleveland, OH 44143. Tel: (800)458-3330. Comprehensive water testing.

Division 12 - Furnishings

Residential furniture is rarely included in the construction contract. The owner will typically select the furniture and have it installed on her or his own or with the guidance of an architect or interior designer. Nevertheless, we are including some guidelines for the selection of healthful furniture in this book about home construction because new furnishings can have a major impact on indoor air quality.

Most standard furniture is built like most standard housing. It is mass produced with little or no thought about the health of the buyer. For those of you who have gone to great effort to create a healthy home, shopping wisely for healthy furnishings is the next logical step. Once again you will find yourself in the role of a pioneer. Most furniture salespeople will not understand what you mean when you speak of healthy furniture. Yet formaldehyde and other chemical levels can soar when new furnishings are brought in to the home. The furniture can continue to pollute the environment throughout its life.

As with the production of building materials, there are many broader environmental concerns pertaining to the manufacture of furniture. These include the use of endangered wood species, toxic waste produced at the manufacturing facility, factory workers' exposure to hazardous chemicals, wasteful packaging, and the exploitation of exporting countries. These factors are discussed in depth in other publications. We will concentrate herein on health concerns related to the homeowner.

Wood Furniture

Most newly constructed wood furniture is actually veneered wood attached to a core of particle board or plywood. These manufactured sheet goods are bound with urea formaldehyde glues which will offgas for many years. Even so-called solid wood pieces may contain hidden plywood or particle board components in order to save on production costs. When selecting wood furnishings, keep the following recommendations in mind.

- Purchase solid wood furniture that does not make use of veneers or sheetgoods. Hardwoods are preferable because they emit fewer terpenes than softer woods. Numerous farmed hardwoods are available. Old growth forest need not be destroyed by virtue of your furniture selection. Although the initial purchase price for solid wood furniture may be more expensive, you will be investing in heirloom quality. Several manufacturers' mail order catalogs are listed in the "Resource List."
- If veneered wood is all that your budget will allow, then consider sealing all surfaces and edges with one of the low VOC vapor barrier sealants listed in "Division 9 - Finishes." In addition, consider other materials such as wrought iron and glass for tables and wicker or rattan for seating. Note: Cane furniture should be clean of mildew and mold. Furniture imported from tropical countries is often sprayed with pesticides while in transit.
- Veneered furniture imported from Denmark is constructed with low emissions sheetgoods to meet that country's more stringent standards.

Finishes on Wood Furniture

Durability, not health, is the criterion used by manufacturers when choosing finishes on wood furniture. The majority of commercial wood sealers are solvent based and will outgas harmful chemicals. We offer the following suggestions:

- Look for furniture with low VOC, water based, natural oil or wax finishes.
- Buy unfinished furniture and refinish with low VOC finishes.
- If you purchase furniture with standard finishes, air it out before placing it in your living space. Finishes will eventually offgas.

Upholstery

Most commercially available upholstered furniture is stuffed with synthetic foam or latex. Many foams will initially have a strong odor. They will break down over time and emit fine particles of chemical dust into the air. Polyurethane foams are extremely hazardous when burned. Furniture stuffing can be made with natural ingredients such as wool, down, kapok, and organic cotton batting. Although these alternatives are not widely available in readymade form, you may find an upholsterer in your vicinity who is willing to work with you. The **Natural Alternative Company** and **Furnature**, Inc. offer natural, organic, upholstered furniture by mail order.

NOTE: Down and kapok stuffings can be allergens for some people.

Upholstery textiles are often synthetic and treated with toxic chemicals to improve stain resistance. Look for natural, untreated upholstery fabrics such as organic cotton, wool, or silk. Selected sources appear in Chart 12.1, "Resource List." **AFM Safe-Choice Lock-Out** (refer to "Division 9 - Finishes") can be used on some fabrics to help repel dirt and stains. Materials must be tested for shrink resistance and color fastness prior to application.

Window Dressings

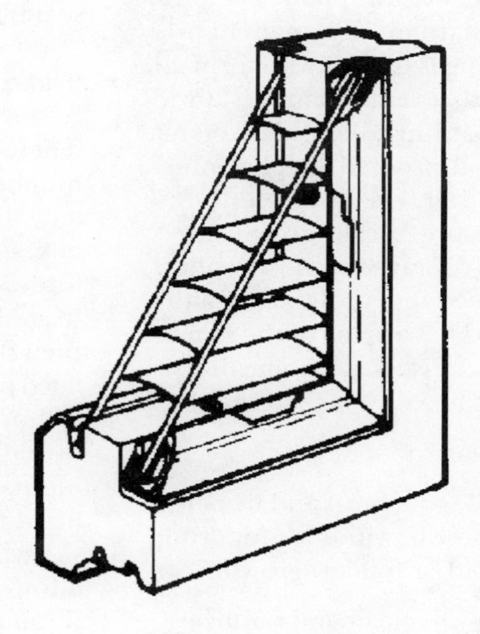

Cutaway view of narrow slat blinds between panes by Pella Corporation.

Most window dressings are made of synthetic fabrics treated with chemicals to make them wrinkle resistant. The recommended drycleaning process further adds to the chemical load they contribute. Natural fabrics can also be problematic because ultraviolet light breaks down the fabric, thereby creating dust and the need for frequent replacements.

Naturally finished wood shutters, louvers, or metallic venetian blinds can be attractive solutions that avoid the problems associated with fabric window dressings.

Pella Corporation produces a line of windows that come with retractable shades sandwiched between double window panes.

Shower Curtains and Liners

New PVC liners and shower curtains have a strong odor from toxins offgassing into the air. Many shower curtains are treated with harmful chemicals to create mildew resistance.

Cotton duck cloth curtains are naturally water repellant, wrinkle resistant, and attractive. They can be machine washed and dried. They are available through several mail order companies including **Seventh Generation** and **Heart of Vermont**.

Beds and Bedding

The most important furniture choice with regard to health is the choice of bedding. We spend approximately a third of our lives in bed. Infants and children spend even more time in bed. While asleep, our noses are in close contact with bedding.

Standard mattresses are made of synthetic fabrics and padding, and treated with petrochemical fire retardants. Permanent press bedding is treated with formaldehyde that remains in the fabric after washing. Wool blankets may be mothproofed with harmful chemicals. Even pure cottons, unless organically grown, are heavily sprayed with pesticides.

A bed that promotes health should have many of the same characteristics as a home that promotes health. The bed should be

- Nontoxic
- Able to absorb and dispel moisture without supporting mold or mildew growth
- Easy to clean and sanitize
- Nonconductive of electricity (free of metal)
- Highly insulative

We recommend the following bedding system which fulfills the abovementioned characteristics. First, the mattress is made of layers, a 1" to 4" thick, untreated, organic cotton futon is topped with a 1" to 3" wool futon. The layers rest on a slatted frame raised above the floor at a comfortable seating height.

The cotton futon provides firm back support while the wool futon, placed on top, adds resilience. Varying thickness and numbers of layers will accommodate different firmness preferences. To properly maintain a futon, it should be aired weekly in sunlight to sanitize it, and then fluffed and replaced in a rotated position so it will wear evenly. A thin futon has an advantage over thicker mattress arrangements because it can be easily lifted and carried.

It is important that air be allowed to circulate under the futon to facilitate evaporation of moisture, thereby preventing mold or mildew growth. A slatted platform will serve to hold the futon firmly in place while providing air circulation around it.

Healthy choices for bedding include organic cotton, silk, or linen sheets; organic cotton flannel sheets; and down, silk, wool, or organic cotton comforters, duvets, and blankets. Some of the specialty mail order suppliers listed in Chart 12-1, "Resource List," carry these items.

Case Study:
The bedroom as sanctuary

J.D. is a 51-year-old man who came to see Dr. Elliott, complaining of insomnia, asthma, and fatigue. After taking an exhaustive environmental history, it became clear that J.D.'s symptoms began during the time he lived downwind from a location where aerial spraying was carried out seasonally for pest control. It appeared that the repeated pesticide exposures had left the patient feeling debilitated, without his usual zest for life and with multiple medical problems, including allergies and sensitivities to a wide range of substances.

As part of his treatment program, J.D. was advised to reduce his exposure to toxins in his home. Since he was on a limited budget owing to his decreased earning capacity, he concentrated his cleanup efforts primarily on the bedroom. At a later date he intended to focus on the rest of the house.

Given the time he spent in bed, J.D. realized that his bed should be the healthiest place in the house. He had recently purchased a mattress made of artificial foam. The synthetic fibers were emitting formaldehyde fumes as the mattress aged which probably contributed to the tight feeling in his chest upon waking. Fortunately, J.D. was able to sell his boxspring mattress and purchase an organic, cotton futon which he placed in an untreated wooden frame. His formaldehyde impregnated, wrinkle free sheets and polyester bedding were exchanged for 100% organic cotton pillows, sheets, and blankets. Because of his concern about possible dust mites in the mattress, he used an organic cotton barrier cloth, woven so tightly that it was impenetrable to these insects. He laundered his bedding frequently in unscented, nonchlorinated detergent.

After J.D. recovered financially from replacing his bedding, his next project was to pull up the old carpet in the bedroom. Although the carpet was several years old and no longer outgassed toxic fumes, it was still a reservoir for dust, dirt, and microorganisms, in spite of frequent vacuuming. J.D. wanted a floor that was attractive, health enhancing, and easy to clean. He chose to install presealed cork flooring because it resembled wood, yet felt soft to the bare foot. He placed two untreated wool scatter rugs on the floor which could be easily taken up and cleaned.

The heating system in J.D.'s house is forced air. The ductwork had been cleaned on a regular basis and electrostatic air filters were used on the return air ducts. Nevertheless, J.D. decided to close off the vents to his bedroom and use an electric ceramic heater. In addition, he bought a portable air filter for the bedroom which contained an HEPA filter for dust, mold spores, and pollens; and a charcoal filter for fumes. The electric motor in the air filter was sealed to avoid toxic emissions and the unit itself was housed in a metal box.

J.D. did not know whether he was sensitive to electromagnetic fields. Since there would be little time or expense involved, he decided to take the necessary measures to reduce the EMFs. He discarded his electric blanket, substituted his digital alarm clock for a battery operated unit, moved his telephone into an adjacent room, and plugged his television into the other side of the room so that the screen was more than eight feet from his head.

The curtains on the windows were replaced with naturally finished wooden louvers which were handsome and easy to clean. The room was cleaned once a week with a simple solution of vinegar and water. He was careful not to introduce toxic odors such as air fresheners, fabric softeners, colognes, and other artificially scented household products. When he needed to occasionally dryclean his clothes, he left them in a closet outside the bedroom. And he was careful to remove his shoes before entering his sanctuary.

J.D.'s efforts paid off. He noted a definite improvement in his overall health. He was now able to get a full night of uninterrupted sleep and awoke feeling refreshed, without the tight sensation in his chest. His energy increased and he was able to think more clearly. J.D. gradually regained his enthusiasm for life and is now trying to convince his friends of the benefits of bedroom sanctuaries.

Discussion

We spend an average of eight hours a day in our bedrooms. Sleep is an important time for rest and recovery for all of us, whether we are sick or in the best of health. Designing our bedrooms with special care can create a healing environment where our bodies can mend from the daily barrage of exposures which we all experience to varying degrees.

Chart 12-1: Resource List

Manufacturer/Catalog	Products	Contact points
Allergy Relief Shop Inc.	Organic cotton mattresses, futons, sheets and blankets.	Allergy Relief Shop; 3371 Whittle Springs Road; Knoxville, TN 37917; (800)626-2810
Bright Futures Futons	Many styles of futons and couch beds.	Bright Futures Futons; 3120 Central SE; Albuquerque, NM 87106; (800)645-4452, (505)268-9738
The Cotton Place	Untreated fabrics.	The Cotton Place; 215 Forest Ave.; Laguna Beach, CA 92561; (800)444-2383, (714)494-3002
Crate and Barrel	Solid wood, glass, and metal furnishings and accessories.	Crate and Barrel; P.O. Box 9059; Wheeling, IL 60060-9059; (800)323-5461
Ecological Beginnings	Solid hardwood nursery furnishings with nontoxic finishes.	Ecological Beginnings; P.O. Box 23698; Chagrin Falls, OH 44023; (440)543-3180
Furnature Inc.	Chemical free upholstered sofas, chairs, and mattresses using 100% organically grown ingredients.	Furnature Inc.; 319 Washington Street; Brighton, MA 02135; (617)782-3169
Heart of Vermont	Bedding and other "products for the chemically sensitive and environmentally concerned."	Heart of Vermont; 131 South Main Street; Barre, VT 05641; (800)639-4123
Homespun Fabrics and Draperies	Handwoven, 100% cotton fabrics without finishes or chemicals.	Homespun Fabrics and Draperies; P.O. Box 3223-NN; Ventura, CA 93006; (800)251-0858
Janice Corporation	Cotton mattresses and bedding.	Janice Corporation; 198 Rt. 46; Budd Lake, NJ 07828; (800)526-4237, (201)691-2979
The Natural Alternative	Upholstered furniture with natural and organic cotton and wool batting, organic cotton canvas upholstery, and colored, washable, organic cotton slip covers.	The Natural Alternative; 11577 124th Street North; Hugo, MN 55038; (612)351-7165
The Natural Bedroom	Formerly Janz Design. Natural bedroom furniture and bedding.	The Natural Bedroom; P.O. Box 2048; Sebastopol, CA 95473; (800)365-6563, (415)920-0790

Manufacturer/Catalog	Products	Contact points
Nigra Enterprises	Air filtration systems.	Kanan Road; Agoura, CA 91301-3328; (818)889-6877
Palmer Bedding	Bedding products.	Palmer Bedding; 9310 Keystone St.; Philadelphia, PA 19114; (214)335-0400
Pella Corporation	Windows come with optional "Slimshade" blinds between the two layers of glass; the blinds never require cleaning.	Pella Corporation; 102 Main Street; Pella, Iowa 50219; (800)547-3552, (515)628-6457
Pottery Barn	Solid wood furniture, glass and metal furniture and accessories, cotton window dressings.	Pottery Barn; P.O. Box 7044; San Francisco, CA 94120-7044; (800)922-5507
Seventh Generation	Organic bedding, shower curtains.	Seventh Generation; 49 Hercules Drive; Colchester, VT 05446-1672; (800)456-1177

Further Reading

Leclair, Kim and David Rousseau. *Environmental by Design.* Hartley and Marks,1992. Overview of larger environmental concerns related to furniture manufacturing.

Division 13 - Special Construction

Swimming Pools

If you wish to include a swimming pool or hot tub inside your home, the two major health concerns to consider are water sterilization and mold infestation.

The standard disinfectants used to kill microbes and algae are chlorine or other halogenated compounds which are easily absorbed through the swimmer's skin as well as inhaled into the lungs. There are several alternatives to chlorination. Ozonation is a popular method used in Europe for sterilizing water. Other methods include electrolysis, ultraviolet light, and filtration through charcoal and pesticide free diatomaceous earth. Pools using these alternate methods need to be frequently monitored for growth. Occasionally a small amount of harsher chemicals may be required.

Enclosing a large body of heated water within a living space will create a humid microclimate which is an invitation for mold growth and can result in damage from condensation. Design measures can prevent mold growth and condensation damage. The following strategies should be integrated into the design.

- Adequate mechanical ventilation and dehumidification
- Fitted cover for the pool or hot tub that remains in place when the pool is unoccupied in order to prevent evaporation
- Watertight enclosure sealed on the inside in order to prevent water damage to the surrounding structure
- Surface finishes impervious to water and easily cleaned
- Rigorous maintenance program to remove condensation and mold growth as soon as it appears

There can be advantages to having a large body of heated water inside the home. The water acts as a reservoir for solar heat storage and humidification. Strategies for taking advantage of these benefits should be considered in the initial design process.

However, due to the intensive upkeep required to maintain a pool or spa so that it does not impact indoor air quality, we do not readily recommend including an enclosed body of water inside a healthy home. We suggest that it might be preferable to locate the pool outside the home.

Case Study:
Asthma from chlorinated swimming pool

B.W. is a five-year-old boy who came with his parents to consult with Dr. Elliott regarding his asthma. The most recent flare-up had occurred during a school field trip to the local swimming pool. Upon further questioning, a pattern emerged that revealed a relationship between water and the triggering of asthma in the child. Dr. Elliott suspected that the chlorine in the water was acting as an irritant to the boy's airways. She suggested to the family that they swim in one of the public pools that had switched to ozone for water purification. In that particular pool, chlorine was used as a supplement, but only in very small quantities. The family was happy to note that their son could now swim comfortably with his friends without difficulty breathing. The family went on to purchase filters for their shower heads which effectively removed chlorine from the showers. They also removed all chlorinated cleaning products from their home. Now that there was one less triggering agent for the asthma, Dr. Elliott could more effectively focus on strengthening the boy's lungs.

Discussion

Chlorine is a poison used to kill bacteria in the water. It is absorbed through skin, inhaled into the lungs, and ingested. At room temperature, chlorine is a gas with a pungent smell. It is very reactive, combining readily with most elements to form compounds, many of which are known to be carcinogenic, such as chloroform, trihalomethanes, and organochlorines.

Symptoms commonly seen from swimming in chlorinated water include runny nose, red eyes, cough, asthma, joint pains, swelling, nausea, urinary discomfort, rashes, and hives. We suggest that you use a less toxic disinfectant for your pool.

Environmental Testing

Diverse quality control tests or checks may be desirable throughout the construction process and selection of materials. This testing can help ensure that materials and installations are as specified. Planning for many of these tests in advance is recommended. Waiting until the last minute will result in costly construction delays since many of these procedures will require the order of test kits, hiring specialists, or waiting for laboratory results.

Material Testing

In choosing healthy materials, you and your architect will base decisions on information supplied by the manufacturer, such as product literature and MSDS sheets, as well as on the appearance and smell of the products. While certain hazardous substances, such as lead, asbestos, and PCBs, are no longer a concern for products manufactured in the United States, precautions may be required if you are using recycled or imported materials. Available tests for lead and asbestos are included in Chart 13, "Resource List." Materials tests you may want to consider are discussed below.

Formaldehyde Testing

This simple spot test is a quick, do-it-yourself check which can be used to ensure that products containing formaldehyde are avoided. Although many manufacturers are now more careful about using formaldehyde than they once were, it is still a common additive in many products. The cumulative effect of several products containing only moderate amounts of formaldehyde in a new home can have severe health consequences. Our approach is to avoid this chemical whenever possible; formaldehyde free substitutes can be located.

The test involves a drop of clear test liquid that changes color in the presence of formaldehyde and other harmful aldehydes. A drop of test solution is placed on the material in question and allowed to stand for two minutes. If the drop changes from clear to purple, formaldehyde is present. The shade of purple can range from a faint pink to a dark plum depending on the concentration of formaldehyde. The test must be read at exactly two minutes because the drop will eventually turn purple even if no aldehydes are present. The solution leaves a purple stain and should be used in a place where it is not visible.

Surface Sampling for Fungus

Materials damaged with mold growth should be rejected. But not all stains are from mold. To determine if mold is a problem, there are three do-it-yourself methods for collecting mold samples. Laboratory analysis will probably be required in each case.

Bulk sampling

With this method a small amount of the material in question is collected into a doubled plastic bag and sent to the laboratory. A teaspoonful of the suspected material is probably enough.

Tape sampling

A piece of clear cellophane tape is pressed onto the surface to be tested. The best place for sampling is at the edge between the stained area and the clean area. The tape is then adhered to a plastic bag or glass slide and shipped to the laboratory. The tape sample is strained by the lab to make the fungal growth easier to view, then examined under a microscope.

Culture collection

A moist, sterile swab is used to swab one square inch of the surface. The swab is inserted into a sterile container and shipped to the laboratory for analysis. Special sterile swabs called "culturettes" are good for this type of sampling. The culturette is presterilized and comes with a fluid-filled glass ampule to provide just the right amount of moisture. The glass ampule and swab are housed in a sterile plastic tube.

Immediately before collecting the sample, the area of the tube over the glass ampule is squeezed to break the ampule and release the fluid which then soaks the cotton swab. The moistened swab is slid from its sterile tube and used to wipe the area to be tested. The swab is then inserted back into the sterile plastic tube for shipment to the lab. Since this method uses liquid, the fungal spores will be hydrated and begin to colonize. It is important to ship the specimen to the lab via overnight delivery service; otherwise the test may be invalid.

The practice of testing for molds using culture dishes is becoming less common. Certain harmful molds such as aspergillus and penicillium are very light and have a tendency not to settle on cultured dishes. They are therefore underrepresented in the analysis.

Other methods of testing for airborne fungal spores and contaminated materials are available but require a trained technician with sophisticated equipment.

Radioactivity

Although radioactivity in building materials is rare, John Banta's home inspections have revealed radioactive stone and tile glazes. Highly radioactive materials can be tested simply by holding a radiation detector next to the material.

For lower levels of radiation, measurements of longer duration should be performed. A minimum of one pound of the material in question is placed in a glass container with an instrument for measuring radioactivity. A useful instrument designed for this purpose is **RadAlert** whose small size allows it to easily fit inside a one gallon glass pickle jar along with the material to be tested. The meter should be set for total counts and left to measure for a timed period of 12 to 24 hours. As a control, the test must also be performed in the same way, in the same location, but with the jar empty. Repeat both tests several times to be sure a radiation emitting solar flare or short-term cosmic event did not interfere with the results. The total number of counts recorded for each test should be divided by the total number of minutes the test ran in order to provide an average count per minute. A substance that measures less than 10% higher than the control test is considered to be free of radiation. Readings with more than a 20% difference from the control test are considered to be significant.

Moisture Testing

Ensuring that materials are dry is essential in healthy building. Building materials can be ruined by moisture damage. Appearing below are four building practices that can be responsible for warpage and deterioration of materials and microbial growth.

- Application of finish flooring materials over insufficiently cured concrete slabs
- Failure to quickly and thoroughly dry out precipitation that enters an unfinished structure
- Installation of wood members with a moisture content greater than 17%
- Enclosure of walls containing wet applied insulation systems, such as cellulose or spray foams, before they are properly cured

It is not always possible to detect if a material is wet by visual inspection. A variety of test procedures have been developed to assist in determining if a material is dry.

Moisture Meters

There are two general types of moisture meters. The first uses sharp pin probes that are pushed into the material to be tested. The pin probe meter detects moisture by electrical conductivity since wet materials conduct greater amounts of electricity than dry materials. This meter leaves pin holes in the materials being tested.

The second type of meter sends a radiowave into the material. The degree of moisture determines how the meter will register the returning wave.

Both types of moisture meters are battery operated and can be used repeatedly. The meters range in cost from about $200 to over a thousand dollars and can require some technical experience. For example, damp wood is measured with a different setting and scale than damp concrete or brick. Companies that specialize in fire and flood damage restoration are likely to have this equipment and be experienced in its use. If you do decide to purchase or borrow a moisture meter, plan on spending some time becoming familiar with it and thoroughly read the owner's manual and instructions. Keep in mind that hidden metals or salt deposits may falsely indicate that materials are wet when in fact they are dry.

Waterproof Testing

All homes are supposed to be weather tight, but many are not. One simple method for testing is to literally water the house. Your architect can specify that the exterior of the house shall be weather tight before any interior construction begins. Once the exterior is complete and the doors and windows are installed and caulked, spray the house with a hose so that every part of the house gets soaked for at least 15 minutes. Then inspect all areas inside the house for leaks. A moisture meter will be useful for this task. This test should only be performed prior to the installation of interior sheathing so that leaks can be easily detected and dried out.

Humidity Meters

Once a structure has been weather sealed, a humidity detector can be used to determine if the interior of the building contains too much moisture. In general, interior moisture levels should always be around 45% or less than outside levels, but this is not always easy to determine without experience. Inexpensive meters for testing temperature and relative humidity can be purchased at most electronics and hardware stores. The relative humidity varies according to the temperature because hot air holds more moisture than cold air.

A special chart called a "psycometric table" can be used to convert relative humidity into absolute humidity. These figures can be used to determine if a structure is dry enough. Certified water loss technicians are available for making these calculations and determinations. If a structure becomes wet during the building process, you should treat it like any water accident (e.g., a leaky roof or a broken pipe), and consult an expert immediately. After a couple of days of delay, the damage can be astronomically expensive. Quick drying by circulating dry heat and air and using mechanical dehumidification is cheaper than removal of mold or rot at a later date.

Calcium Chloride Moisture Testing

Large quantities of water are present in cement, gypcrete, aircrete, and other poured masonry materials. These materials must be adequately dried before finishes are applied. It is common in new construction for a carpet or other floor finish to be laid on a slab before the slab is thoroughly dry. Further drying is inhibited, allowing microbial spore levels to climb as mold invades these damp areas.

A kit for testing moisture in masonry has been developed which uses calcium chloride salts. These salts absorb moisture from the air at a known rate. The kit contains a plate with the calcium chloride salt, a plastic dome, and an adhesive material. The calcium chloride is weighed to the nearest hundredth of a gram. Scales for measuring the salts can be found at your local pharmacy. The calcium chloride is then placed in the area to be measured and covered with the plastic dome which is sealed to the slab with the adhesive material. After two days the cover is removed and the calcium chloride reweighed. Based on the weight gain and the number of hours that have passed, a determination of the area's moisture content can be made. The instructions for the kit also contain a chart that will help determine when the slab is dry enough to proceed with various finishing materials.

Energy Efficiency and Air Flow Testing

Blower Doors

Blower doors consist of a sophisticated fan set in an adjustable frame. They are used to test air flow and pressure in a home. There are many uses for blower doors, such as detection of leaks in the walls and in HVAC ductwork. You can also determine if the ventilation is adequate, as well as identify the location of energy leaks in the structure.

The equipment requires extensive training to use. We recommend that you hire a technician to carry out blower door testing. For most new homes this testing will cost several hundred dollars. Dollars saved in energy conservation from identified and corrected leaks will offset the cost of testing over a few years.

Theatrical Fog Machine

Certain parts of the home, such as garages, attics, and crawl spaces should be completely sealed from the rest of the house in order to prevent the passage of contaminated air into living spaces. One easy way to test for leaking air flow is by using a theatrical fog machine. This is the same equipment used on stage and in movies to create fog for special effects. The equipment can be rented from most theatrical supply companies. The unit is placed in the area to be tested and turned on to fill the space with fog. The adjoining areas are observed for signs of the fog, and leaks can be easily pinpointed and sealed.

When testing the garage, the door and open vents need to be sealed with tape and plastic to prevent the fog from escaping. The same is true of attic and crawl space vents and intentional openings to the outdoors. Common air infiltration points revealed by the fog test include electrical outlets, the juncture where the sheet rock meets the floor, and around poorly sealed plumbing, electrical, and ductwork penetrations.

Be sure to notify the fire department before you begin this type of test; otherwise a well-meaning neighbor who sees the smoke might dial 911 and set the fire trucks in motion.

Radon Testing

In Division 7 we discuss radon gas and mitigation. Several varieties of acceptable methods are currently being used to measure radon in the air. Some test kits are available through local hardware stores. It is important to precisely follow the manufacturer's instructions. General principles that should be followed regardless of the type of kit used are listed below.

- Close the home for a minimum of 12 hours before beginning the test and keep it closed throughout the testing period. Entry and exit from the house are permissible as long as the doors are not left open.
- Place the sampler about 30 inches above the floor and at least two feet away from the wall in the area being tested. Keep the sampler away from doors, windows, fireplaces, outside walls, and corners, and any other places where drafts or stagnant air may exist. These precautions are necessary to ensure that the tester is exposed to a representative sample of air.
- Verify that the starting and stopping time are accurately recorded. This information, along with the date, must be included with the tester when it is returned to the lab. Without precise recording information, the results cannot be considered valid.

A typical radon test kit costs less than $25. After each individual test, the kit must be returned to a laboratory for analysis. Multiple testing or continuous monitoring can be carried out with electronic radon monitors.

Radon Soil Testing

Radon mitigation is most effective and least costly when incorporated into the construction of the home. If you are building a new home and there is reason to suspect a radon problem, then a soil test is advisable. Although the test will not provide definitive results as to what the radon levels will ultimately be in the finished home, it will nevertheless be an indicator that will help you decide whether mitigation measures should be included in your construction plans.

The test kit available for measuring radon in the soil involves placing a special collection box with its open side over the soil to be tested. Soil is mounded around the lip of the box to form a tight seal and keep the box in place. Radon gas is then trapped and concentrated in a carbon medium which can then be measured by a testing apparatus. The starting time and date are recorded. After the prescribed period of time, usually 48 hours, the soil is pushed away and the tester retrieved and returned to its foil pouch. The stop time is recorded and sent with the other information and materials to the lab for analysis.

Testing for Chemical Fumes

New homeowners are often assaulted by a barrage of chemical smells, and even though they might not consider themselves to be chemically sensitive, they may be bothered or made ill by the fumes in their homes. Sniffing finishing materials such as upholstery, carpets, and paint before they are installed will reveal important information. However, even if a building product or material passes the sniff test when sampled, the odor can become unbearable once the product is installed in the home. This is because chemical fumes accumulate inside the house and are emitted from a much larger surface area than that of the sample.

If you are unsure about how you will tolerate a product once it is applied or installed in your house, we recommend that you test the product before purchase in a manner that will simulate the level of concentration in the home. One method of testing is to place a sample of the product in question in a large, glass jar with the top screwed on tightly to allow fumes to accumulate. The following day the jar is opened and the contents are sniffed for unacceptable fumes. If the sample is too large to be placed inside a container, it can be placed next to your pillow while sleeping. It is important that the sample be new. For example, a carpet swatch that has been in a showroom for three years will not provide an accurate indication of what a freshly unrolled carpet will smell like in your home.

This type of testing, although somewhat helpful, has obvious limitations. While the test gives information about the particular product in question, it does not indicate cumulative nor synergistic effects when combined over time with other chemicals. Since you cannot predict these effects in advance, the goal is to choose products with the least amount of odor or toxic emissions.

Chart 13-1: Resource List

Manufacturer	Product	Contact points
Swimming Pool Purification Systems		
Aqua Zone	Water ozonation system that eliminates chloramines and kills bacteria. Can reduce the need for harsh chemicals by as much as 70%.	Aqua Zone; 79 Bond St.; Elk Grove Village, IL 60007-1298; (847)439-4454
Environmentality World	Water purification systems for pools and spas. DC current is passed across electrodes, causing formation of oxygen radicals which are lethal to microorganisms.	Environmentality World; 221 Pawnee, Suite 207; San Marcos, CA 92069; (800)288-0230
Real Goods	Source of "Floatron," a solar-powered pool purifier, combining solar electric power with mineral ionization. Reduces chlorine usage up to 80%. Also source of Trifield meter for detecting electric, magnetic, and microwave pollution.	Real Goods; 555 Leslie Street; Ukiah, CA 95482-5576; (800)762-7325
Test Kits		
AirChek Inc.	Sells a variety of home test kits for radon, lead, formaldehyde, water purity, and microwave leakage.	AirChek Inc.; Naples, NC 28760; (800)247-2435
The Allergy Relief Shop	Mail order catalog with microbiology lab for culturing mold and other microorganisms.	The Allergy Relief Shop; 3371 Whittle Springs Road; Knoxville, TN 37917; (800)626-2810
Environmental Health Center	Easy to use mold testing kits for the home. $35 for kit and analysis.	Environmental Health Center; 8345 Walnut Hill Lane, Suite 220; Dallas, TX 75231; (214)373-5149
Mold Survey Service	Mold test.	Dr. Sherry Rogers; P.O. Box 2716; Syracuse, NY 13220; (315)488-2856
Spotcheck Pesticide Testing Kit	Instant pesticide checking kit based on technology developed by the U.S. military applications. Ships with supplies to carry out four tests of soil, water, food or surfaces. Some of the pesticides that can be tested include Sevin, Dursban, Diazinon, Malathion, and Parathion.	The Cutting Edge Catalog; P.O. Box 5034; Southhampton, NY 11969; (800)497-9516
Testing Consultants		
Ozark Environmental Services	For air and water testing, and consultation regarding toxic gases, molds, asbestos, volatile organic compounds (VOCs), pesticides, gas leaks, EMFs, and radon.	Ozark Environmental Services; 114 Spring Street; Sulphur Springs, AK 72768; (800)835-8908
Rad Alert	Device for measuring radioactivity.	International Med Com; 7497 Kennedy Rd.; Sebastopol, CA 95472; (707)823-0336

Manufacturer	Product	Contact points
ReCon Restoration Consultants	Years of experience with biological contamination of indoor environments, and restoration after fire and water damage.	ReCon Restoration Consultants; 3336 Sierra Oaks Dr.; Sacramento, CA 95864; (916)736-1100
Streit's Healthy House Service	EMF assessments and solutions for the home. Manufacturer of remote circuit shut-off switch.	Streit's Healthy House Service; 2726 Corrales Road; Corrales, NM 87048; (505)898-8432

Division 14 - Conveying Systems

This division is not used in residential construction.

Division 15 - Mechanical

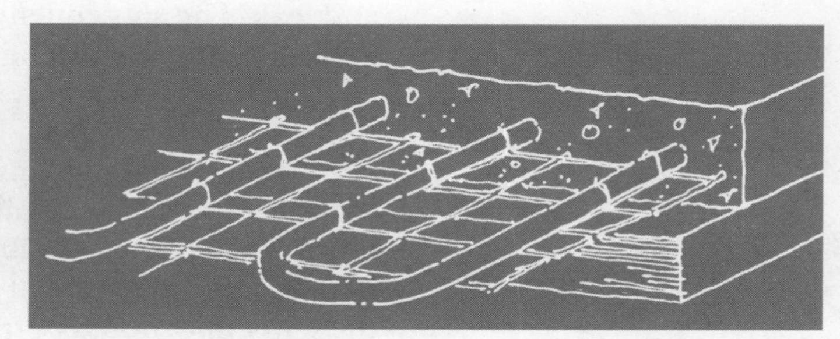

Water Supply and Waste

Polyvinyl chloride (PVC)is the standard for residential supply and waste piping. PVC plastic piping has been shown to outgas diethyl phthalate, trimethylhexane, aliphatic hydrocarbons, and other harmful gases. It should not be used for water supply piping in a healthy home. We recommend seeking alternatives for waste lines as well because of the pollution resulting from both the manufacture and disposal of PVC piping. Consider specifying the following items for a healthier plumbing installation.

Water Supply

Supply Pipe

Acceptable alternatives for supply pipe are listed below.

- Type L Copper: Type M copper contains lead and is not recommended. Solder shall be lead free, silver solder. The system shall be flushed to eliminate any flux from the soldering operation prior to occupancy.
- **Aquapex**: A cross linked polyethylene that shall be installed by certified installer.

Pipe Assembly

- Assemble pipes with the longest pieces possible to minimize the amount of glue or solder required.
- When possible, glue pipe assembly outside the building envelope.
- Wipe up excessive glues and protect all surfaces from glue drips and spills.

Waste Drain System

- Waste drains shall be ABS piping.

Floor Drains

Appliances containing water, such as water heaters and washing machines, can malfunction and leak. The subsequent water damage and mold can be avoided if you plan for this possibility. By providing strategically located floor drains or drain pans, the water from accidental spills can be diverted to the sewer line or to the outdoors. Drains that lead to the sewer line should be installed with a trap to prevent unwanted sewer gases from entering the home. It is important that the traps be kept filled with water, or "primed," which creates a physical barrier against the entry of sewer gases.

Plumbing Penetrations

Where plumbing penetrates through walls and ceilings, the air space created around them must be completely sealed to prevent unwanted air infiltration. Consider specifying the following item.

Penetrations

Wherever plumbing penetrates the wall, apply 100% silicone caulking to create an airtight seal.

Backflow Protection

In some communities sewage systems periodically back up and flow into homes leading to devastating contamination. Devices installed at the point where the home waste line meets the municipal sewage line are available to prevent this. The local planning department may be able to help you determine if backflow prevention devices are advisable. In many communities, claims for sewage damage will not be paid unless such devices were in place prior to the incident.

Residential Heating and Cooling

Methods of heating, cooling, and ventilating homes have many important health ramifications that will impact our lives long after the initial building materials have outgassed and reached a neutral state. Ideally, if we lived in a pristine natural environment with low humidity and mild temperatures, we would be able to condition our homes without mechanical assistance by means of solar gain, shading, and cross ventilation. Residents throughout most of the United States do not have such luxury. Cold and cloudy winters, hot and humid summers, and polluted or pollen filled air are realities from which homes must shelter occupants.

We have come to expect a level of comfort and temperature control in our homes undreamed of by our not too distant ancestors.

Along with the increased comfort level, we have also unwittingly come to accept many health problems associated with heating and cooling systems. In fact, to a greater extent than with any other building system or component, heating and cooling methods can be a major cause of sick building syndrome. Some of these problems are listed below.

- Toxic fumes from gas, oil, or propane fuels that work their way into the building envelope through leaky supply lines and improperly sealed mechanical rooms, and from open combustion appliances
- Backdrafting of hazardous and sometimes fatal gases into the living space from the chimney
- Infiltration of pollutants from outdoors due to depressurization
- Fried dust resulting from hot surface temperatures
- Circulation of dust through convection type heating
- Leakage of toxic substances used in heating appliances or refrigerant gas from cooling appliances
- Contamination by mold growing in the ductwork and air conditioning equipment
- Fiberglass fibers originating in ductwork insulation which circulate in the living space

Later in this chapter we present options for heating and cooling that include guidelines for healthier installations, language for specifications, and maintenance suggestions that will help to eliminate some of the problems mentioned above. In the next section, we focus on ways of reducing the need for mechanical heating and cooling.

Reduction of Heating and Cooling Loads through Design Strategies

The application of a few simple design and planning principles can greatly reduce the mechanical heating and cooling loads required to live comfortably, thereby improving health and lowering energy consumption. In designing the home for energy efficiency, consider the following suggestions.

- Choose an exterior wall system with a high insulation value.
- Choose interior wall and floor systems with high levels of thermal mass to assist in retaining heat in the winter and cold in the summer.
- Seal cracks and joints to prevent unwanted infiltration and exfiltration.
- Choose a high insulation value for the ceiling. This measure will be especially cost effective because most heat escapes through the roof.

Consider the surrounding site as an extension of your climate control design.

- Make use of deciduous trees to shade in the summer and allow solar gain in the winter.

- Observe prevailing wind patterns in planning for natural ventilation.
- Consider trees as windbreaks to lower the heating load created by cold winter winds.
- Situate your home as far away from pollution sources as possible so that the site can provide a quality air supply for home ventilation.

Take advantage of solar heat.

- Orient the home to take advantage of solar gain.
- Plan fenestration (arrangement of doors and windows) for the desired amount of heat gain.
- Make use of overhangs and sun angle information to prevent overheating in the summer.
- Provide natural cross ventilation to facilitate natural air exchange and to provide natural cooling in the summer.
- Use light colors to reflect heat and dark colors to absorb and store heat.
- Provide thermal mass for heat storage.
- Use thermal window shading devices to control heat loss.

Become a more active participant in temperature control.

- Open and close windows to control fresh air and temperature.
- Open and close thermal shading devices to control heat gain and loss.
- Be willing to add and subtract layers of clothing to allow for a greater range of acceptable temperatures.
- Consciously temper your body to acclimatize to a broader comfort range.

Straw bale exterior walls provide high R-values while adobe interior walls store heat produced by a solar greenhouse. (Architect: Paula Baker. Contractor: Living Structures, Inc. Photo: Paula Baker.)

Healthier Heating and Cooling

Each heating and cooling system has advantages and disadvantages that must be carefully weighed when choosing a system that best fits your needs and budget. Once you have made a choice, there are several considerations in design, construction, and maintenance that will optimize performance and minimize the health risks of the system. In the design phase, certain factors must be considered by you and your architect from inception, such as the location of the mechanical room. During the construction phase, the choice of materials and installation procedures can influence the ultimate outcome. For this reason we have provided specifications for the contractor where relevant. Finally, a regular cleaning and maintenance program is essential for optimal efficiency. This task will ultimately fall to the owner and may influence your choice of HVAC system.

Choice of Fuel Source

Gas and other sources of combustion fuels can pollute the air stream if you do not plan carefully. Electric heat is often considered "cleaner" heat because combustion does not occur in the home. However, environmental pollution from electricity generation plants must be acknowledged. Moreover, electric heating appliances generate electromagnetic fields, an invisible and often overlooked source of pollution. For many homeowners the higher cost of electric heat makes it unaffordable. Whatever your choice of fuel source, there are several strategies that can be employed in the mechanical room that will make heating healthier.

Mechanical Room Heating Options

- The mechanical room should be a dedicated room, insulated and isolated from the living space either by creating a separate building to house the equipment, or by creating a well-sealed room that opens to the outside.
- The equipment in the mechanical room may produce elevated levels of electromagnetic fields (EMFs) and should not be located adjacent to heavily occupied living spaces.
- Ensure the supply of adequate combustion air to the mechanical room.
- A fire alarm located in the mechanical room is recommended.

Heating and Cooling Appliances

We recommend the following guidelines for choosing, locating, and maintaining heating equipment.

- Purchase equipment designed for backdraft prevention.
- We recommend sealed combustion units to prevent transfer of combustion by-products into the air stream. This is especially important where the mechanical room must be accessed directly from the living space.
- If you are using a forced air system, we strongly recommend adding a good combination filtration system which will filter out both particulate matter and gas.
- If possible, choose a heating system which does not run hot enough to fry dust. Hydronic systems and heat pumps used in mild climates meet this requirement.
- Institute a regular maintenance program to clean components, purge mold or mildew growth, and change filters at specified intervals.

Heating and Cooling Options

Hydronic Heating

Hydronic heating, delivered through hot water, is usually a wall mounted baseboard or radiant floor system. Baseboard systems are usually made of copper tubes and aluminum radiating fins with painted steel covers. Baseboard radiators can be noisy if not maintained, and they can collect dust and dirt.

Radiant floor systems are usually made of plastic, rubber, or copper tubing installed within or under the floor system. Radiant systems are silent and clean.

The advantages of a hydronic system include slightly lower operating costs, even heating, quieter operation, ease of zonation, and independent room temperature control. Disadvantages of the hydronic system include slow response time and slightly higher maintenance costs compared to forced air, due to the number of mechanical components. We suggest the following specifications for a hydronic system.

Hydronic Heating

- Insulate all piping located in non-living areas.
- Some baseboard units are subject to offgassing when the paint on covers overheats. Verify if this is a problem prior to purchase.
- The aluminum fins on baseboard units tend to collect dust and should be cleaned regularly.
- Some infloor systems use very odorous rubber products. While this is not a problem where they are embedded in concrete, it can be a source of indoor pollution where the tubing is exposed at access points. We recommend **Wirsbo Hepex**, an odorless, cross linked polyethylene tubing for radiant floor heating.
- Avoid using metal piping which can conduct electromagnetic fields through the structure.

Forced Air

Forced air heating delivery systems are generally sheet metal ducts that supply air to each room. Advantages of a forced air system include the ability to control humidity, filter air, introduce fresh air from the outside, and quickly change the temperature of an area. Disadvantages of the forced air system may include greater operating costs, noisy operation, distribution of unwanted fumes and particulate matter, larger space requirements for equipment installation and housing, and the necessity to modify standard systems to prevent unwanted dehumidification. Forced air heating, which heats the air, is considered to be far less comfortable than radiant heating which heats objects.

If forced air is your choice for heating and cooling, then you can take advantage of the whole house air distribution ducting that will already be in place to improve air quality by implementing the steps below.

- Use a fresh air intake vent from the outside to the furnace in order to introduce and distribute fresh, tempered, ventilated air into your home. Be careful to avoid locating the vent in proximity to places such as trash storage areas or where auto exhaust and other pollutants could be brought inside the house.
- Install enhanced filtration into your forced air stream. (See the "Air Filtration" section below.)
- Avoid the use of plastic flex ducts. They are difficult to keep clean and easily damaged.
- Choose a furnace with sealed combustion to avoid the entry of combustion by products in the air stream.

Case Study:
A constant supply of warm dust

A retired couple contacted John Banta because they were experiencing eye irritation and difficulty breathing due to dust in their home. In spite of frequent vacuuming and dusting, an unusually heavy deposit of dust was noted on the furnishings during the house inspection. John suspected that the furnace system was the source of contamination because the heat registers in the home were lined with a fine dust and the clients' symptoms worsened when the furnace was on.

John was puzzled, though, by the lack of dirt on the cold air return filter and the absence of air movement. He opened the cold air return and examined the inside wall to see if there were any visible obstructions. To his surprise, he found no duct at all. The cold air return was a dummy and went nowhere.

Further investigation revealed that the furnace and duct system were located in the crawl space under the home. John inspected the crawl space where he discovered that there was no connection between the cold air return port on the furnace and the rest of the house. In fact, the furnace was taking cold air from the crawl space and blowing the unfiltered, contaminated air directly into the house. Consultation with a heating and air conditioning company was recommended to correct this construction defect.

Discussion

HVAC duct systems should always be leak tested to ensure that they meet specified standards. The stated industry standard for a sealed duct system is less than 3% leakage, which is rarely achieved. The furnace itself will account for much of the leakage since it is difficult to seal. The furnace should be mounted in a clean, easily accessible area like a mechanical room, and not in an attic or crawl space.

Another cause of leakage is from unsealed joints where the metal ducts fit together. Since the return side of the furnace is sucking air back into the furnace, it will suck contaminants through leaks in the ductwork. If the unsealed ducts pass through walls or attics containing fiberglass, then these fiberglass particles are sucked into the ducts and blown into the house. If the unsealed ducts are in a crawl space under the home, then moldy, pesticide laden, or dusty air can be sucked into the furnace system and blown into the house.

Forced Air Ductwork

Care must be taken during the design, installation, and maintenance of forced air ductwork as it is often the source of allergies and other health problems associated with forced air heating and cooling.

A large amount of dust and debris is generated during the construction process which frequently finds its way into the ductwork and then becomes a source of air contamination once the system is in operation. Floor registers should ideally be avoided because debris will inevitably accumulate in them, not only during construction but in the course of occupancy as well.

Another common source of air contamination in HVAC systems is the use of ceilings, joisted floors, and wall cavities as ductless air plenums. These unsealed plenums frequently act as pathways for contaminated attic or crawl space air to enter into the building. Fibers from wall and ceiling insulation are frequently sucked into the heating system and circulated throughout the building envelope. Furthermore, the plenums are inaccessible for cleaning and impossible to seal.

In addition to the abovementioned concerns, consider the following characteristics if you use forced air ductwork.

- Ductwork and sealer should be made from nontoxic materials
- Ductwork should be leak proof
- Easy access to ductwork is necessary for future inspection and maintenance
- Ductwork should be insulated where necessary to prevent heat loss and condensation

In order to achieve an optimal ductwork system, we suggest the specifications listed below.

Ductwork Installation

Ductwork shall be made of sheet metal and provided with cleaning and inspection openings.

- Ductwork shall be well-sealed with nontoxic compounds or tapes. Mastics will be water resistant, and water based, with a flame spread rating no higher than 25, and a maximum smoke developed rating of 50.
- During construction, the ends of any partially installed ductwork shall be sealed with plastic and duct tape to avoid the introduction of dust and debris from construction.
- All forced air must be ducted. The use of unducted plenum space for the transport of supply or conditioned air is prohibited.
- **RCD6** is a nontoxic, water based mastic recommended for sealing ductwork and metal joints.
- Do not use cloth duct tape. It has a high failure rate that can result in undetected leakage.
- Seal all joints, including premanufactured joints and longitudinal seams.
- Gaps greater than 1/4" must be reinforced with fiber mesh.
- Do not use below slab ductwork. It will collect moisture and dirt, providing a breeding ground for microbes.
- All ductwork running through uninsulated spaces shall be insulated to a minimum of R10 to prevent condensation problems and to save energy.
- Duct insulation must be on the outside of the ductwork. Ductwork insulated on the inside can release harmful fibers into the air stream.
- Locate supply and return registers on walls or ceilings, not on the floor, to help prevent debris and dust from falling into the ducts.
- Ductwork must be professionally cleaned prior to occupancy.

Once the ducts are in place, a regular maintenance program is essential to maintaining a healthy system. Identify a professional maintenance company that uses high-powered duct cleaning equipment. Avoid the use of chemical cleaners.

Combined Heating and Cooling Systems

Heat pumps are far more energy efficient than electric resistance heat and can be used for both heating and cooling. Heat pumps extract heat from outside air or, in some cases, from a water source. Air source heat pumps are most common in areas where winter temperatures seldom fall below 30 degrees and where summer cooling loads are high. As temperatures fall below 30 degrees, the heat pump must rely on electric resistance heating to make up the difference, at which point it loses its economic advantage.

Advantages of the heat pump follow: a single unit carries out both heating and cooling needs, humidity is not added to the air, and operation is quiet. These units are most cost-effective and energy conserving in regions with moderate climates where heating and cooling loads are not extreme.

Cooling Systems

Common types of air conditioners include condensing or refrigerated air conditioners, electric heat pumps as discussed above, and evaporative coolers.

Condensing air conditioners are available either in small units designed to cool one area of a home, or central air conditioners which will cool an entire home via ductwork. Advantages of central air conditioners are out-of-the-way location, quiet operation, introduction of less pollen into the indoor air, integration with the forced air heating system, and greater cooling capacity and efficiency than portable models. However, these systems are expensive to operate and consume a lot of energy. They cost up to seven times more to operate than evaporative cooling systems.

It is important to choose an AC unit which continues to blow air across the cooling coils for a time after the cooler is turned off. This allows any moisture remaining on the coils to be dried off, thereby discouraging mold growth.

Room air conditioners are less expensive to install than central air conditioners. Since they only cool designated areas, they save money and energy. However, they do tend to be noisy.

Evaporative coolers are practical in very dry areas and are available either as a "direct model," which adds humidity to the home, or an "indirect model" which does not add humidity.

Evaporative cooler operating costs are significantly lower than for condensing units, and evaporative units are fairly inexpensive to install. Fresh outdoor air is brought into the living space and stale air is exhausted. Evaporative coolers have a lower cooling capacity and work well only in low humidity conditions, such as in the West and Southwest.

When using mechanical air conditioning, you can save energy and money by keeping the windows closed. One exception to this rule is in the case of evaporative coolers which are more efficient when windows are left open. Air conditioners should be shut off and windows opened at night when it is cool outside. Do not cool unoccupied rooms or homes. Insulate all exterior ducting. This can save you at least 10% of the energy costs of cooling.

Another name for evaporative coolers is "swamp" coolers. They must be kept clean or they truly become swamps filled with microorganisms. Maintain systems regularly, keeping coils and filters clean. Locate the cooler in a shaded area.

Chart 15-1: Heating and Cooling Systems Summary

Type of system	How it works	Advantages	Disadvantages	Comments
Heating Systems				
Forced air	A fan pulls air through a heating unit and distributes the air throughout the house via ducts.	•Can be easily adapted for filtration, humidification, dehumidification. •Almost immediate response time. •Inexpensive to operate.	•Less comfortable than radiant heat. •Stirs up and fries dust. •Can exacerbate allergies. •Ductwork is architecturally cumbersome. •Leaky ducts can depressurize home. •Noisy. •Needs regular cleaning. •Metal ductwork grounds negative ions. •Fumes from gas or oil fuel can enter air stream. •Insulation particles can enter air stream.	•Many of the disadvantages of forced air can be rectified by adding filtration to the system both at the furnace and where the air exits into the room.
Radiant hydronic floor heat	Hot water is run through plastic or metal tubing in floor or under floor. Natural convection gently distributes heat.	•Even, comfortable heating. •Comfortable at lower temperatures. •Efficient. •Not hot enough to fry dust. •Silent. •Low maintenance. •Easy zonation. •Invisible.	•Slow response time. •Initial installation costly. •Does not filter air. •Not practical for cooling.	•Avoid metal tubing which can transmit EMFs.
Liquid filled base-board heaters	Hot liquid is circulated through fin tube base-board units and radi-ates into room.	•Heats quickly. •Comfortable radiant heat. •Not hot enough to fry dust. •Less expensive than in-floor heating.	•Baseboard units are dust traps. •Limits furniture placement. •Can be hot to touch.	•Heated surfaces of baseboard units may offgas. •Leaks (other than water) may be toxic.
Electric radiant floor, wall, or ceil-ing heat	Electric current passes through resistant wir-ing imbedded in walls, floors, or ceilings.	•Even heating. •Comfortable radiant heat.	•Expensive to run. •Can create high levels of EMFs. •Less expensive systems run hotter and fry dust.	•Not recommended in a healthy home because of EMFs and high degree of energy consumption.
Electric baseboard heating	Individual units are plugged in.	•Initial installation is inexpensive and easy. •Does not require centralized machinery. •Puts heat only where required.	•Expensive to run. •Hot to touch. •Traps and fries dust. •Emits EMFs.	•Heated surfaces may offgas.

Type of system	How it works	Advantages	Disadvantages	Comments
Wood burning stoves	Wood fire is contained in a noncombustible stove. Heat radiates into room.	•Radiant heat source. •No central equipment required. •Inexpensive to install and operate.	•Messy to run and requires high maintenance. •Reports show much higher rate of respiratory problems in children in homes using wood stoves. •Burn and fire hazard. •Chimney can be subject to backdrafting. •Burning wood produces more than 200 toxic by-products of combustion. •Most of the heat escapes up the chimney.	•Not recommended in a healthy home. •Choose the most efficient models available, burn hardwoods, and clean flue often.
Masonry heater ("Kacheloffen")	Heat from wood fire travels through a series of masonry chambers, is stored in the masonry mass, and slowly radiates into the room.	•Very efficient use of fuel requires less tending than conventional wood stoves. •Burns cleaner. •Produces comfortable radiant heat that does not fry dust or burn people. •Inexpensive to operate, requires no further equipment. •Can incorporate cook stove or oven. •Can be an architectural feature.	•Initial installation is costly; few craftspeople in the United States know how to build. •Generates a small amount of combustion by-products.	•Less convenient than central heating systems. •Considered the most healthful way to heat according to baubiology.
Passive solar heating	Heat from the sun is captured through glazing and stored in building components with high thermal mass such as concrete and adobe walls and floors.	•No operation expenses. •Does not consume fossil fuels. •Does not fry or circulate dust.	•Dependent on the weather. •Requires a relatively high degree of human interaction. •Must be incorporated into architecture.	•For more information refer to "Further Reading" section.
Heat pump	Heat or cold is extracted from outside air and transferred to inside air.	•Can be used for heating or cooling. •Cost effective in mild climate. •Quiet.	•Not cost effective where temperatures are frequently below 30 degrees. •Uses freon as transfer medium.	
Cooling Systems				
Central refrigerant coolers	Freon gas is passed through a condenser. Heat is transferred to the outdoors and the cool air is distributed throughout house via ductwork.	•Can also dehumidify air. •Will handle large cooling load. •Can be quiet to operate if condenser is remote. •Shares ductwork with central heating.	•Expensive to operate. •High energy consumption. •Uses freon. (Freon is an atmospheric ozone depleter. •Requires maintenance to prevent mold.	•Drip pan must be inspected and cleaned regularly for mold free operation.

Type of system	How it works	Advantages	Disadvantages	Comments
Room refrigerant coolers	Freon gas is passed through a condenser. Heat is transferred to the outdoors and the cooled air is blown into the room.	•Inexpensive initial installation. •Because cools only designated, occupied areas, energy waste and expense are reduced.	•High consumption of energy. •Uses freon. •Requires maintenance to prevent mold. •Noisy.	
Evaporative (swamp) coolers	Air is passed over a wet medium. As evaporation occurs air is cooled and then blown into home.	•Low cost initially and when in operation. •Uses no CFCs or HCFCs. •Requires 80% less energy than refrigerant coolers. •Works well in hot, dry climates.	•Subject to mold and other microorganism growth. •Not suitable in humid conditions. •Cannot take as large a load as refrigerant models. •Can be noisy. •Requires maintenance to keep mold free and needs frost protection in cold winter climates.	•Should be drained, cleaned, and dried at the end of summer use period.

Ventilation

Until the 1960s, ventilation in homes occurred naturally, obviating the need for intentional ventilation systems. Homes were loosely built, allowing enough outside air to make its way through the home to keep it fresh. By some accounts, this loose construction contributed to as much as three to four air exchanges per hour. Currently, with energy efficient construction, much of the unintentional air exchange has been eliminated. Whereas in the past, homes were built of more natural, nonpolluting materials, in recent years indoor air has become at least five to ten times more polluted than outdoor air, and often too polluted for optimal health. Although minimum air exchange rates are enforced for commercial structures, this is not the case for residential construction, except where exhaust fans are mandated.

Ventilation, like many other components essential to health, is considered an "extra." ASHRAE (The American Society for Heating, Refrigeration, and Air Conditioning Engineers) has set a standard of .35 air exchanges per hour and 15 cubic feet per minute (cfm) per resident for residential ventilation. Although this may be sufficient to dispel pollutants created by human activity, it will not be enough to dispel the chemical pollution generated by standard construction, or the thousands of other chemicals introduced into homes through furnishings, clothing, cleaning products, cosmetics, and other scented products. ASHRAE determines its requirements based on the level at which 80% of a test population feels comfortable. It should be noted that it is quite possible to feel comfortable in environments that are polluted enough to be detrimental to health. The human body has the ability to become accustomed to harmful chemicals, much like one might adapt over time to the toxic effects of tobacco smoke.

Ventilation strategies are necessary even in a healthy home because, with tight construction, air exchange is necessary to ensure fresh air and dispel odors from everyday living. Care should be taken to locate the fresh air supply away from exhaust air piping and in the most advantageous ocation for receiving an unpolluted air stream.

Chart 15-2: Residential Ventilation Strategies

Type	Purpose	How it works	Advantages	Disadvantages	Comments
Natural ventilation	To bring fresh air into the home and exhaust stale air.	Takes advantage of natural air patterns. Strategically placed openings encourage fresh air to move diagonally through the space, entering low and exiting high.	Quiet. Free. Maintenance free. Does not require energy to operate.	Can be drafty and create greater heating/cooling load. Air cannot be filtered. Allows for minimal control.	This strategy works best in mild climates and can be enhanced through various roof ventilation techniques.
Exhaust fans	To remove localized pollution at the point of generation. Primarily used for kitchens and baths.	Stale and moisture laden air is sucked out of house at the point of generation, using a powerful fan.	Pollution is quickly removed before the rest of the home is affected.	Can depressurize home causing infiltration and possible backdrafting.	It is important to supply replacement air when fans are in operation. Exhaust fans are required by code in bathrooms and laundry rooms without operable windows and in kitchens.
Supply fans	To provide fresh air.	A fan blows fresh outside air into the home creating positive pressurization which forces stale air out.	Inexpensive. Pressurization of home prevents contaminants from infiltrating from outdoors.	Cold drafts around fan in winter. Pressurization can cause hidden moisture problems as humid air is forced through wall openings and then condenses.	Adequate and strategically placed vents to exhaust air are required. This is a good strategy for venting a basement area. Supply fans are not suitable for dispelling kitchen and bath generated pollution.

Type	Purpose	How it works	Advantages	Disadvantages	Comments
Balanced mechanical ventilation	To provide fresh air and exhaust stale air while controlling pressurization.	A set of fans brings fresh air in through intake and distributes it, then exhausts stale air to the exterior.	Provides balanced pressurization. Comes equipped with, or can be adapted for, various filtration strategies.	Does not moderate temperature or humidity of incoming air. Can be noisy. Relatively small fans are standard and are insufficient to handle large amounts of gas filtration.	
Air to air heat exchange or HRV (heat recovery ventilator)	As above. Also moderates the temperature of fresh supply air.	Incoming fresh air passes through a series of chambers adjacent to outgoing exhaust air. Heat, but not air, is transferred from one to the other.	Reduces heating and cooling costs by recovering 60 to 80% of heat.	Chambers can be made of paper that collect dirt or plastic which can offgas. Choose one with metal chambers. More costly initially than other balanced ventilators. Causes condensation; must be maintained to remain mold free.	Most effective for tight homes in cold climates.
Fresh air intake incorporated into forced air system	To provide fresh air when the central forced air system is operating.	A 3" to 6" metal pipe with damper valve provides fresh air into the furnace supply stream.	Inexpensive to retrofit. Ventilation supply air is preheated or precooled. Makes use of existing ductwork for distribution. Creates slight positive pressurization and can compensate for air lost through leaky ducts.	Only operates during heating or cooling season. Depends on a well-maintained heating and cooling system to deliver good quality air.	Screen all intake pipes to prevent rodent infestation.

Air Filtration

The addition of filters to ventilation and forced air heating and cooling systems allows for greater control of air quality. As discussed above, indoor air is often too polluted to properly nourish occupants. The first line of defense against such pollutants is to provide an ample supply of fresh outdoor air through ventilation. Unfortunately, "fresh" air, although considerably cleaner in most cases than indoor air, contains allergens in the form of molds and pollens and manufactured pollutants, including exhaust fumes, smoke, and pesticides. When your immediate surroundings are less than perfect, you may wish to incorporate some form of filtration into your home.

Home ventilation systems can easily be adapted to filter large particles like pollen and mold spores. However, most home ventilation systems are not designed with very powerful fans and therefore cannot handle the air resistance created by some of the more efficient filtration methods, especially those designed to remove gases. Consequently, whole house filtration is often more successfully combined with the forced air distribution system. Standard filters used with most forced air systems are designed primarily to prevent large particles from harming the motor, and are insufficient to effectively filter out small particles injurious to human health. Most forced air equipment must be adapted to receive additional filtration systems. A forced air system, when equipped with good filters, will not only clean fresh intake air, but will continue to clean air as it recirculates.

Chart 15-3: Residential Filtration Strategies

Filter type	Purpose	How it works	Efficiency*	Advantages	Disadvantages	Comments
Standard furnace filters	To filter out large particulate matter to safeguard the motor, not inhabitants	A coarse, 1" thick filter traps large particles.	5% of particulate matter	Inexpensive. Easy to change.	Indoor air quality is not significantly improved.	Can easily be replaced with 1" media filter which will raise efficiency to 20%.
Medium efficiency extended surface filter	Particulate filter	Air is strained through a pleated filter with an extended surface area that maintains air flow.	40%-50% of particulate matter	Relatively inexpensive. Sufficient for most general filtration. Air flow resistance can be low enough to use with ventilator.	Filtration is inadequate for very polluted environments and/or very sensitive people. Does not filter out gaseous pollution.	Special adaptation required to work with HVAC. Media filters become more efficient with time as pores become smaller, but then air resistance increases.
HEPA (high efficiency particulate air filter)	Particulate filter	Polyester or fiberglass fibers are bound with synthetic resins, creating a medium with extremely small pores.	97%+ of particulate matter	Can remove minute particles for extremely clean air. Can remove cigarette smoke.	High air flow resistance requires powerful fan. Expensive. May require custom design. Does not filter out gaseous pollution such as VOCs.	Not commonly used in residential filtration. A carbon post filter will eliminate odor generated by HEPA filter. An inexpensive, frequently changed pre-filter will extend life of HEPA filter.

Filter type	Purpose	How it works	Efficiency*	Advantages	Disadvantages	Comments
Electrostatic precipitator (ionizer)	Particulate precipitator	Mechanism is mounted to ductwork which statically charges dust. Dust is collected at oppositely charged plates in a filter.	90% of particulate matter when clean	No resistance to air flow. Efficient when clean. Does not require replacement.	Must be adapted for residential use. Ozone is produced as by-product of high voltage. Relatively expensive. Does not filter out gaseous pollution. Only efficient when clean. Generates EMFs.	Plates must be cleaned regularly.
Electrostatic air filter	Particulate filter	Electrostatic charge is generated by friction as air moves through special media.	10 to 15% of particulate matter	Good for mold and pollen. No customization required on some filters used with HVAC. Inexpensive.	Not efficient for capturing small particles. Limited efficiency. Does not filter out gaseous pollution.	May be substituted for standard furnace filters. An inexpensive way to bring relief from pollen and mold allergies.
TFP (turbulent flow precipitator)	Particulate precipitator	Turbulent air stream "drops" particles into collection space where there is no air flow.	Manufacturer claims 100% for particulate matter	No resistance to air flow. Can be used with ventilator. Very low maintenance.	Product is new on the residential market and does not have an established performance record. Does not filter gaseous pollution.	
Partial bypass filter	Absorption of gaseous pollutants.	Granules of absorptive material are held in place and separated by a metallic grid. Some air passes through medium and some flows past unrestricted.		Allows some air to flow through, thereby cutting down air resistance and requiring less powerful fan.	Not suitable where air is highly polluted. Not suitable in ventilator.	Works in conjunction with HVAC where the same air is repeatedly run through the filter.
Activated carbon filter	Absorption of gaseous pollutants.	Gases cling to many faceted carbon granules.	Varies	Effectively removes gases with high molecular weight. Offered in standard furnace sizes for low pollution situations.	Does not remove certain lightweight pollutants such as formaldehyde or carbon monoxide. Filters become contaminated with use and can release pollutants if not changed.	Can be treated to remove more gases. Must be changed regularly per manufacturer's recommendations.

Filter type	Purpose	How it works	Efficiency*	Advantages	Disadvantages	Comments
Activated alumina	Adsorption and transformation of gaseous pollutants.	Activated alumina is impregnated with potassium permanganate. It acts as a catalyst in changing the chemical composition of harmful gases and also acts through adsorption.		Will remove gases not removed by carbon, including formaldehyde. Lasts longer than carbon.	Not as absorptive as carbon. More expensive than carbon.	Activated alumina changes color when depleted.

* Efficiency particulate ratings are based on research by John Bower in *Understanding Ventilation: How to Design, Select and Install Residential Ventilation Systems* (Bloomington, IN: Healthy House Institute, 1995).

Chart 15-4: Resource List

Product	Description	Manufacturer/Distributor
Aquapex	A cross linked polyethylene nontoxic plumbing system.	Wirsbo; 5925 148th Street W.; Apple Valley MN 55124-9928; (800)321-4739
Hepex	Cross linked polyethelene tubing for radiant floor heating.	Same
RCD6	Nontoxic water based mastic for sealing ductwork and metal joints.	Positive Energy; P.O. Box 6578; Boulder, CO 80206; (800)488-4340; (303)444-4340
Sources for Portable Air Filtration and Heating Products		
Air filters	High quality air filters with HEPA and charcoal.	Allermed Corporation; 31 Steel Road; Wylie, TX 75098; (972)442-4898
Same	High quality air filters.	E.L. Foust Company, Inc.; P.O. Box 105; Elmhurst, IL 60126; (800)225-9549
Diverse systems	Broker for environmentally benign equipment and systems (e.g., air and water filtration, heaters, and vacuum cleaners). Free consultation.	Nigra Enterprises; 5699 Kanan Road; Agoura, CA 91301-3328; (818)889-6877
Radiant heater	Long wave, ceramic radiant heater designed for the chemically sensitive. No outgassing from heating elements, and no fans or blowers. Heats room quickly. Does not dry out the air.	Ceramic Radiant Heater Corp.; P.O. Box 60; Greenport, NY 11944; (800)331-6408
Van EE	Air to air heat exchanger.	Shelter Supply; 1325 East 79th Street; Bloomington, MN 55425; (800)762-8399, (612)854-4266

Further Reading

Bower, John. *Understanding Ventilation: How to Design, Select and Install Residential Ventilation Systems.* Bloomington, IN: The Healthy House Institute, 1995.

Mazria, Edward. *The Passive Solar Energy Book.* Rodale Press, 1979.

Division 16 - Electrical

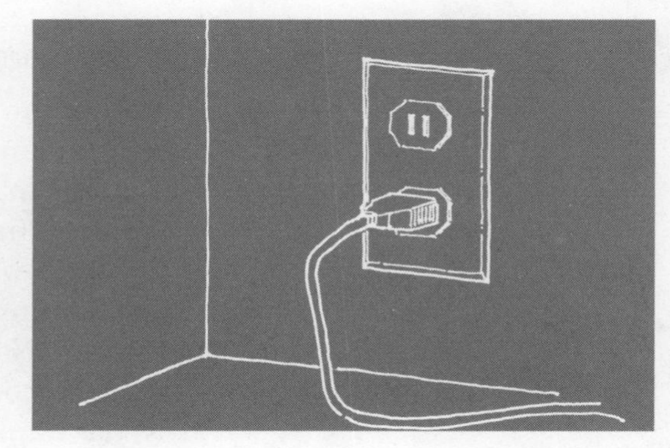

Electric and magnetic fields are commonly discussed as if they were a single entity termed electromagnetic fields or EMFs. In fact, the two phenomena, although interrelated, are distinctly different and each will be discussed separately in this division. The following chart presents a comparison of the two.

Chart 16-1: Electric fields versus magnetic fields

Electric fields	Magnetic fields
Flow in straight lines in all directions from source unless conductors attract them.	Radiate out from the source, flowing in loops.
Can be easily shielded.	Difficult and expensive to shield. (Even lead is not effective.)
Attracted by conductors such as metal or salt water bodies, including people.	Penetrates all normal building materials.
Present when switches for machinery are off or on.	Only occur when appliances are switched on and current is flowing.
Not widely recognized as a health threat in conventional circles at the time of this writing.	Safe exposure limits not regulated by the U.S. government. Sweden has set safe exposure limits.
Reportedly affects the nervous system and can cause insomnia, anxiety, depression, and aggressive behavior.	Reportedly affects cellular function and has been statistically linked in some studies with increased cancer cell growth rate, Alzheimer's, miscarriage, and birth defects. Some sensitive individuals report physical reactions when in elevated magnetic fields.
Electrical code permits but does not mandate reduced electric field wiring.	Electrical code offers protection against exposure to magnetic fields produced by wiring in the structure.
Proper use of electric field meters requires expertise.	Easily measured with a gaussmeter.

Magnetic Fields

Basic Home Wiring and Net Current

Although the relationship between human health and elevated fields remains controversial, there are definite safety concerns associated with wiring techniques that cause magnetic fields. The National Electric Code, in recognition of such hazards, has mandated safer wiring.

Your electrician may be puzzled if you declare that you want a home free of all elevated magnetic fields. However, if you say that you wish to have a home in compliance with the electrical code and therefore free of net current, you are saying in effect the same thing in a language that is universal to electricians.

Most household wiring consists of 110 volt lines. If one were to peel back the outer insulating plastic on a piece of Romex, the most common wiring used, three strands would be revealed — one black, one white, and a third that is either green or bare copper. The black strand is referred to as the "hot" wire because it draws electricity from the breaker box or panel and delivers it to light fixtures and appliances. The white wire, called the neutral, returns the electricity to the panel after it is used. The green or bare copper wire is the ground wire. Under normal conditions, it does not carry electricity. However, if a malfunction such as a short occurs, it serves as a fail-safe protective device by carrying power back to the ground until the breaker is tripped and the power to the faulty circuit is cut off, thereby preventing shock and electrocution.

When the electrical system is functioning as it should, the amount of electricity flowing out to the appliance through the hot wire is equal to the amount of electricity flowing back through the neutral wire. This equal and opposite flow of current through the wires creates a net current of zero which is the desired condition. When, for various reasons, a net current is present, a magnetic field is created.

A second condition which creates net current with associated magnetic fields occurs when the neutral and hot wires are separated by distance. When Romex wiring is used, the hot and neutral wires run adjacent to one another inside the plastic insulating sheathing and cancel each other out.

In an older wiring system known as "knob and tube," the hot and neutral wires were run on separate studs. There was also no grounding. Although now prohibited by code, this dangerous system of wiring, along with its associated elevated magnetic fields, is still found in many older homes.

The National Electrical Code prohibits the production of net current which should protect people from elevated magnetic fields as well. Unfortunately, subtle code violations resulting in the production of net current frequently occur, causing not only elevated magnetic fields, but also increased risk of fire and electrocution.

Case Study:
Magnetic field caused by wiring errors

John was called to the home of a client who was concerned about the high magnetic field in the apartment she was renting. The living room, dining room, and kitchen showed a reading of around 16 milligauss. After carefully tracing the wires, John discovered the problem. The apartment had two light switches by the front door. One switch controlled the outdoor lights; and the other, the living room lights. The two switches were controlled by different circuit breakers. When the switches were wired into the box, the neutral wires were joined together with a single electrical connector nut, a situation known in the trade as "ganged neutrals." The problem was easily remedied with the simple addition of a 14-cent electrical nut which separated the two neutral wires. The magnetic fields throughout the house dropped to 0.5 milligauss, considered to be an acceptable level.

Discussion

This case study illustrates a simple code violation which went unnoticed by the electrical inspector. Had the inspector used a gaussmeter, it would have been easily detected before final closeout. Surprisingly, this is not common practice. Had the tenant not used a gaussmeter, the code violation may never have been revealed.

We have identified several commonly used wiring techniques which create very high magnetic fields. Although these techniques are code violations because they create net current, they often go unnoticed by code inspectors, as happened in the previous case study. The following instructions for wiring techniques should be specified in order to avoid and detect conditions that create elevated magnetic fields in wiring.

Prohibited Practices and Testing for Wiring

- All wiring shall be performed in strict accordance with the National Electric Code.
- The ganging of neutral wires from different circuits is prohibited.
- Edison circuits are prohibited. (Edison circuits occur when three-wire Romex is used to create two 110 volt circuits and the single neutral wire is shared.)
- Bonding screws shall be removed from the neutral bus of all subpanels per manufacturer's instructions.
- When wiring a 1/2 switch outlet using two separate breakers for each half of the outlet, the two neutral wires must not make electrical contact. This is accomplished by breaking off the prescored conductive tabs between the two sections of the outlet per manufacturer's instructions.
- Neutral wires on 1/2 switched outlets shall not be mixed. They shall remain paired with corresponding hot wire.
- All wiring entering an electrical box must be from the same circuit.
- At the time of the final electrical close-out, and in the presence of the general contractor, architect, or owner, the electrician shall apply a minimum load of three amps to the distal end of each electrical circuit. The home shall be inspected under load using a gaussmeter. Any elevated magnetic fields greater than .5 milligauss will indicate the presence of net current.
- It is the responsibility of the electrical contractor to locate and eliminate net current.

Magnetic Fields from Three- and Four-Way Switches

Lights switched from two different locations are called three-way switches. When lights are switched from three or more locations, they are called four-way switches. A correctly wired three- or four-way switch will not emit magnetic fields. However, these switches are often wired incorrectly and thus become a source of magnetic fields that can radiate throughout the entire room. To avoid improperly wired three- or four-way switches, specify the following items.

Three- and Four-way Switches

- Three-wire Romex shall be used between the switches when wiring a three-way switch (see next illustration). If alternate wire is used, it shall be twisted.
- Each three- or four-way switch must be controlled by a single breaker.
- All wiring for three- or four-way switches shall be contained in a single run of wire or a single metal conduit. All runs not in a conduit must be bundled.

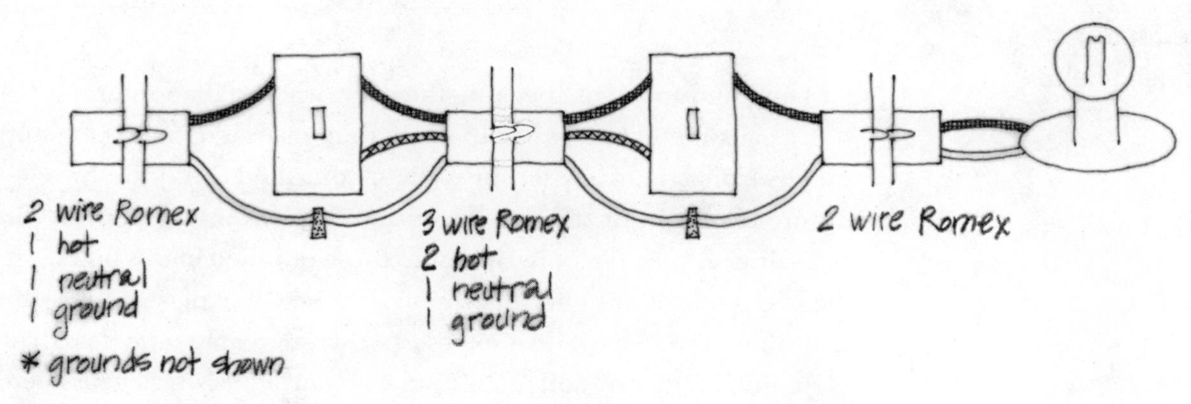

Properly wired three-way switch.

Dimmer Switches

Dimmer switches are a source of magnetic and radio frequency fields. We do not recommend them. If they are used, they should be located at a distance from seating and sleeping areas. The most expensive name brand dimmers tend to emit smaller fields. Choose a model that emits no fields when all the way on or all the way off.

Magnetic Fields in Panels and Subpanels

Many electrical panels and subpanels emit substantially elevated magnetic fields. This problem arises because breaker and neutral bus bars are configured such that the neutral and hot wires are separated once fastened in place. As discussed earlier, this separation causes magnetic fields.

Some electrical panels are configured with the neutral bus bar split to run alongside the breakers. To cancel the fields, the hot and neutral wires would be the same length and installed beside one another. We recommend that such reduced field configuration panels and wiring be specified as indicated below.

Panels and Subpanels

- Panels and subpanels shall be configured so that hot and neutral field cancellation are possible.
- The following panels and subpanels are acceptable: **Siemens EQIII**, standard load center electrical panels and subpanels with split neutral.
- Hot and neutral wires from the same run are to be installed adjacent to one another.
- Wire lengths shall be equal.

Case Study:
Magnetic fields

The importance of checking the electrical installation under load with a gaussmeter before occupancy is demonstrated in this case study. An electromagnetically sensitive client consulted with John by telephone throughout the construction of her home which was built according to specifications similar to those outlined in this book. After the client moved into her new home, she began experiencing symptoms that occur when she is exposed to elevated magnetic fields, such as ringing in the ears and inability to concentrate. Using a gaussmeter, she discovered that about half of the home registered over 5.0 milligauss. She called John in a state of panic, convinced that her house was ruined and that she would never be able to live in it.

John contacted the client's electrician and offered to help him diagnose the problem over the telephone. Under John's guidance, the electrician conducted field testing with the client's gaussmeter. From the measurements, it became clear to John that the problem was located in the subpanel controlling a section of the house. At that point, the electrician immediately realized what he had forgotten to do. Some panels and subpanels are interchangeable except for a single screw that must be removed from the neutral bus bar to electrically isolate it from the ground wires in the panel. Called a "bonding screw," it was causing net current in all circuits in the subpanel. The electrician simply removed the bond screw and the magnetic fields dropped in an instant to less that 0.2 milligauss.

Dielectric Unions

A dielectric union is a plastic joint that acts as an insulator, preventing the passage of electricity between conductive materials. In a typical home, conductive gas and water lines come into contact with appliances in several places. For example, water lines feed into refrigerators with ice makers, and gas lines feed into motorized furnaces. Should a fault occur in the appliance, wayward electricity will be distributed through the piping unless a dielectric union is used to isolate the appliance from the utility pipes. "Electrified" piping is undesirable for the reasons listed below.

- Magnetic fields will radiate out from the pipes.

- Net current in gas lines is an explosion hazard.
- Pipes carrying net current can become an electrocution hazard.
- Electric current flowing through pipes causes electrolysis which results in decomposition of the pipes.

The installation of dielectric unions is an inexpensive safeguard against a rarely occurring phenomenon but one with potentially devastating results. We recommend specifying dielectric unions in healthy homes.

Dielectric Unions

Metallic gas and water lines shall have dielectric unions installed wherever they enter into contact with any electrical appliance, when it is permitted by code.

Magnetic Fields Entering the Home from an Outside Source

As discussed in "Division 2 - Site Work," it is important to choose a site that is free from elevated magnetic fields generated from overhead power lines. Magnetic fields caused by faulty wiring in a neighbor's home can also be transferred into your home through utility service lines. Because electricity will follow all available paths, metal plumbing, gas lines, cable TV lines, and telephone lines can become pathways for uninvited net current. Consequently, taking simple precautions to prevent such an occurrence is prudent when site conditions allow. Although the electrical code mandates grounding and bonding, it does not dictate the configuration of utilities entering residential structures. By grouping the entry point of all utilities and providing proper bonding, any net current traveling through public utility lines will be shunted back without ever entering the home. However, pathways of elevated magnetic fields may be created in your yard. These too can be blocked, but will require the expertise of a knowledgeable consultant.

Grouping the entry point of all utilities along with the proper bonding and grounding will also provide more protection against lightning damage. However, this is not a substitute for lightning rods, which are designed to take a direct lightning strike to the home. If site conditions do not allow for the grouping of all utilities, then testing with a gaussmeter for unwanted fields with the house power turned off would be a prudent safety precaution, both during construction and periodically thereafter.

Appearing below are specifications for preventing the entry of magnetic fields through utility services.

Bonding and Grounding

- All utilities, including telephone, cable TV, gas, and water shall enter the building at approximately the same location, within a 4-ft radius. If site conditions do not permit this ideal configuration, and if new magnetic fields are detected throughout the structure before it is energized, then there is reasonable cause to suspect that fields are entering from an outside source. At this point, an expert must be consulted to properly block the unwanted fields. Because neighborhood conditions may change over time, fields should be checked regularly.
- All utilities entering the structure shall be properly bonded immediately prior to entry.
- Bonds or grounds shall occur at only one point along each utility.

Case Study:
Net current in utilities

After purchasing a gaussmeter, an electrician was surprised to discover an elevated magnetic field throughout his entire driveway and a portion of his home. Upon learning that the field did not decline when he shut off the power to his home at the main breaker, he concluded that the source of the field was from net current in the gas line. A gas company technician visited the site and confirmed that the gas line was carrying electricity. There was no cause for concern, he said, because the amount of electricity was small. The electrician was not comforted by such reassurances.

As a specialist in complex wiring techniques for boats and marinas, he was familiar with the problems of electrolysis and galvanic action resulting from electricity straying from its intended path. In fact, he had even witnessed boats at the local marina whose metal had gradually dissolved from exposure to net current. Thus, the electrician reasoned, the net current in his plumbing and gas lines would cause the lines to deteriorate at an accelerated rate. After informing a gas company representative that the galvanic action in the pipes was a liability for the company due to the possibility of an explosion, the electrician was finally able to persuade the company to take his complaint seriously.

Measuring Magnetic Fields

Magnetic fields are measured with a gaussmeter. A homeowner might consider purchasing a gaussmeter for one or more of the reasons listed below.

- To determine safe distances from various household appliances.

- To help detect wiring errors which not only produce magnetic fields, but may also be fire and electrocution hazards.
- As a periodic safety check to determine that no new problems have developed in household appliances.
- As a periodic safety check to ensure that no new magnetic fields are entering the home through utility lines.

There are two basic types of gaussmeters: single and triple axis. Single axis meters tend to be less expensive, and are slightly more difficult for a novice to use because they must be rotated to align with the flow of the magnetic field in order to detect it. Pointing a triple axis meter at a field in order to detect it is not necessary because this meter requires positioning only within the range of the field. Less expensive gaussmeters will give false readings when measuring certain magnetic fields such as those generated by computers and electrically ballasted fluorescent lights. The following table provides a comparison of widely available gaussmeters that are adequate for measuring household fields.

Chart 16-2: Gaussmeters

Brand name	Axis	Approximate cost	Accuracy for TV, fluorescents, computers
Dr. Gauss	Single	Under $20	Overestimates 60 Hz field due to interference from higher frequency fields
Tri-Field	Triple	$150	Overestimates 60 Hz field due to interference from higher frequency fields
MSI Magcheck	Single	$200	Accurate for 60 Hz field; does not measure higher frequency fields
Bell 4080	Triple	$200	Accurate for 60 Hz field; does not measure higher frequency fields
Tracer 3D	Single/Triple	$1,000	Accurate for 60 Hz field; can indicate the presence or absence of higher frequency fields

Electric Fields

Wiring to Reduce Electric Fields

German baubiologists have long been concerned about the negative health effects associated with exposure to electric fields. In the United States, mainstream science has given little credence to the notion that electric fields pose a health threat. However, we may soon see a change in the prevailing wisdom.

A recent study by the Ontario Hydroelectric Company indicated a sevenfold increase in cancers among workers exposed simultaneously to magnetic and electric fields.[1] The study suggests that the presence of electric fields potentiates the health impact of magnetic fields. These relatively new findings may shed light on why various studies of the impact of magnetic fields on humans have been inconclusive.

A certain proportion of the population appears to suffer from hypersensitivity to electric fields. These individuals react to exposure with immediate neurological symptoms such as insomnia, depression, and anxiety. One frequently reported symptom is that of feeling physically exhausted but too jittery to sleep, or "wired and tired."

Wiring for reduced electric fields is not a requirement by the electrical code and can be costly. Electric fields generated by wiring can be shielded in conduit. This practice is standard in commercial construction, but rarely found in residential construction. Even if conduit is used, electric fields will still be emitted from appliances or fixtures once they are plugged in, unless the unit has been specially wired or renovated. For people with hypersensitivity to electric fields, special wiring may be a necessary expense. Techniques for this type of specialty wiring are beyond the scope of this book and will require consultation with an expert.

The following instructions may be specified to reduce electric fields generated by household wiring.

Wiring for Electric Field Reduction

- All household wiring shall be placed in MX, MC, or rigid metal conduit.
- All electrical boxes and bushings shall be metal in order to provide shielding of electrical fields throughout the entire run to the panel.

A less expensive approach would be to shield the wiring only in areas where occupants spend a great deal of time. A modified shielding plan might be specified as follows.

Wiring for Electric Field Reduction

- All wiring to bedrooms, dens, family room, living room, and dining room shall be in MX, MC, or rigid metal conduit.
- All other electrical runs shall be routed so as to avoid the abovementioned rooms to the greatest extent practical.
- Avoid running wire under bed placement locations.

A third and even less expensive approach is to reduce electric fields exclusively in the bedroom by employing a convenient remote control device which cuts the power to the bedroom at the breaker box. Power can be turned off from the bed just before retiring at night so that the bedroom becomes a field free sanctuary. Because the presence of high electric fields is most commonly associated with sleep disturbances, we believe that such a device is an important feature in electric field reduction for the healthy home. A remote control system is available through **Streit's Healthy House Services**.

Remotes are most effectively used when wire runs are planned in advance. In brief, certain wiring, such as the wires leading to smoke detectors or refrigerators, should not be included with bedroom runs.

Case Study: Electric fields and insomnia

Several years ago John was requested to investigate the house of a woman who claimed that she had not slept well since she moved into the house. Upon inspection of the bedroom, John noted that the electrical fields registered over 5,000 millivolts on the meter. He explained that the goal for a healthy house is 20 millivolts or less. (These measurements are relative, and are measured in the body using special equipment and techniques.)

The elevated electric fields were a result of the electrical wiring in and near the bedroom. The fields were being concentrated in the metallic bedsprings which acted as an antenna, redirecting the electric field upward toward the client. John explained to the couple that the easiest way for an electrician to lower the electric fields in the bed would be to install a remote controlled switch on three of the circuit breakers in the basement that controlled the electrical electrical wiring in and around the bedroom. At that point in the conversation, the client's husband expressed his skepticism regarding the investigation and findings. He doubted that the electrical field could explain his wife's sleeplessness since he did not experience similar symptoms. He was reluctant to follow John's recommendations.

John then suggested that the couple try an experiment to ensure that a remote switch would indeed be money well spent. They were instructed to turn off the three breakers in the basement every night before they went to bed to determine if the woman slept better. John reminded them that there would be no power in the bedroom and that they should have a battery operated alarm clock and flashlight available.

A few weeks later the client contacted John to report that she was sleeping soundly for the first time in years and that both she and her husband were elated. She related to John what had transpired after he left the couple's home. When the time came to turn off the breakers on the first night, she could hear her husband grumbling with resentment and stomping loudly down the steps to the basement to turn off the breakers. That night she slept so long and soundly that she barely made it to the bathroom in time to empty her bladder the following morning. Her husband took note of her improvement and the second night went into the basement to shut off the breakers without saying a word. Again she slept soundly and awoke with the sun, feeling refreshed. By the third night she began to feel romantic, a feeling she had not experienced in a long time. By the fourth night her husband was whistling while he took the basement stairs two at a time. At this point the couple was eager to invest in a remote switching device.

Discussion

Because of standard wiring practices, readings of 1,000 millivolts or higher in a home are typical. Wiring homes for low electric fields is much easier and more cost effective when this consideration is part of the initial building plans. Wiring paths, for example, can be situated in a way that limits the number of circuits involved, and high field emitters can be placed at a safe distance from the sleeping area. Existing homes cannot always be controlled by simply shutting off the breakers. Sometimes expensive shielding is needed.

As is often the case, sensitivity to electric fields varies from person to person. In the case described above, the client developed severe insomnia while her husband experienced no ill effects.

Shielding Electrical Fields Emitted from Refrigerators

Because refrigerators generate large electric fields, we recommend that they be given a dedicated circuit and that the wiring be shielded with one of the recommended metal conduits in order to block the fields. In addition, the metal refrigeration cabinet should be bonded to the electrical ground. Note that the compressor motor and defroster will still produce high magnetic fields. Consequently, the home should be designed so that the refrigerator is at least eight to 10 feet distant from living and sleeping areas.

Residential Lighting

Residential lighting types are typically incandescent, fluorescent, and low voltage. Appearing below are selected pointers concerning residential lighting and EMFs.

- Transformers of low voltage lighting produce a magnetic field. If low voltage lighting is used, then choose remote transformers and locate them in closets at a distance from where occupants spend a lot of time.
- The ballasts in fluorescent lighting emit magnetic fields which may not be detectable on an inexpensive gaussmeter. Fluorescent lighting should be avoided in areas where occupants spend a lot of time, and should never be located on a ceiling below a bedroom.
- If using recessed can lighting, then specify airtight insulation contact (ATIC) cans. These types of cans save energy and prevent dust and attic gasses from filtering into the cans.
- If wiring is run through a metal conduit, then the metal housing of the fixture must be in electrical contact with the metal conduit in order to shield the occupied space from electric fields.

Smoke Detectors

The two basic types of smoke detectors are ionizing and photoelectric. The ionization type contains a radioactive substance called Americium 241. Although the radioactive substance is shielded, we cannot recommend this type because there is no safe place for disposal once the smoke detector is discarded.

Smoke detectors are available for use with 9 volt batteries or for hardwiring into the 110 volt household wiring, with or without battery backup. We recommend a hardwired system with battery backup.

If you are wiring so that your bedroom circuitry can be shut off, it is important to put the smoke detector on a separate circuit so that it will always remain active. If this circuit is run through a metal conduit, the electric field will be minimal.

Sources for Photoelectric Smoke Detectors

- **BRK 2002**: Line of 120 volt photoelectric smoke detectors with battery backup.
- **Natural Choice Item #10222**: 120 volt photoelectric smoke detector with battery backup.
- **Ecoworks**: Three models of photoelectric smoke detectors (9 volt, 12 volt with 9 volt battery backup, and 120 volt).
- **Real Goods 57-122 Smoke Alarm**.

Chart 16-3: Resource List

Product/Service	Description	Manufacturer/Service provider
57-122 Smoke Alarm	Photoelectric smoke alarm.	Real Goods; 555 Leslie Street; Ukiah, CA 95482-5576; (800)762-7325
BRK 2002 Line	Line of photoelectric smoke detectors with battery backup	BRK Electronics; 780 McClure Road; Aurora, IL 60504-2495; (800)392-1395; (630)851-7330
Dr. Gauss	Gaussmeter	Huntar Company Inc.; 473 Littlefield Ave.; San Francisco, CA 94080
EMF reduction supplies	Various	Healthy Home Center; 1403-A Cleveland Street; Clearwater, FL 34615; (813)447-4457
MSI Magcheck and Bell 4080	Gaussmeters	Magnetic Sciences International; HCR2, Box 850-295; Tucson, AZ 85735; (800)749-9873, (520)822-2355
Natural Choice Item #10222	Photoelectric smoke detector with battery backup	Eco Design/The Natural Choice; 1365 Rufina Circle; Santa Fe, NM 87505; (800)621-2591, (505)438-3448
Photoelectric smoke detectors	Same	Ecoworks; 2326 Pickwick; Baltimore, MD 21207; (410)448-3317; (800)466-9320
Tracer 3D and other EMF measurement devices	Gaussmeter	Radiation Technology; 600 N. Hometown Road; Akron, OH 44333; (216)666-7710
Remote circuit shut-off switch.	Radio signal from remote control activates transceiver which in turn sends signal along house power lines to circuit box.	Streit's Healthy House Service; 2726 Corrales Road; Corrales, NM 87048; (505)898-8432
Siemens EQIII	Standard load center electrical panels and subpanels with split neutral	Siemens; 2880 Sunrise Blvd.; Rancho Cordova, CA 95742; (800)964-4114 Widely distributed, available in many home improvement store chains throughout the U.S., such as Home Depot and Builder's Square
Tri-Field Meter	Gaussmeter	Alpha Labs; 1272 Alameda Ave.; Salt Lake City, UT 84102; (801)532-6604
EMF Consultants		
EMF measurements and mitigation consultation, including household electric and magnetic fields, radio and microwave assessment.	Same	Environmental Training and Technology; 1106 Second Street #102; Encinitas, CA 92024; (760)436-5990
Provides list of independent EMF consultants.	Same	National Electromagnetic Field Testing Association; 628-B Library Place; Evanston, IL 60201; (847)475-3696

Chart 16-3: Resource List

Product/Service	Description	Manufacturer/Service provider
EMF detection and remediation.	Same	Restoration Consultants; 3336 Sierra Oaks Drive; Sacramento, CA. 95864; (916)736-1100
EMF detection and energy balancing for the home. Diverse techniques include dowsing.	Same	Streit's Healthy House Service; 2726 Corrales Road; Corrales, NM 87048; (505)898-8432

Endnote

1. *Microwave News* 15:4 (July-August 1996), 1, 5-7.

Further Reading

Becker, Robert O. *Cross Currents: The Promise of Electromedicine, The Perils of Electropollution.* J.P. Tarcher, 1991. A timely and eloquent warning on the hazards of electronic pollution.

von Pohl, Gustav Freiherr. *Earth Currents, Causative Factor of Cancer and Other Diseases.* Stuttgart, Germany: Frech-Verlag, 1987.

Riley, Karl. *Tracing EMFs in Building Wiring and Grounding.* Tucson, AZ: Magnetic Sciences International, 1995. (800)749-9873.

Appendix A.

MCS: A Description and Testimonial

Multiple Chemical Sensitivity or Environmental Illness

Multiple chemical sensitivity (MCS), often referred to as environmental illness, is an immune and nervous system disorder which involves severe reactions to many everyday chemicals and products. For some people, MCS occurs with dramatic onset, precipitated by a major chemical exposure or industrial accident. But for most people, the condition develops gradually as the result of the cumulative exposures of daily life.

The symptoms of MCS are diverse and unique to each person and can involve any organ of the body. Symptoms range from mild to disabling, and sometimes they can be life threatening. These symptoms include headaches, fatigue, sleep disturbances, depression, panic attacks, emotional outbursts, difficulty concentrating, short-term memory loss, dizziness, heart palpitations, diarrhea, constipation, shortness of breath, asthma, rashes, flu-like symptoms, and seizures. Symptoms may be chronic or occur only when a person is exposed to certain substances. The particular organs affected depend on the individual's genetic background and prior history, as well as the specific chemicals involved in the exposure.

Symptoms are often triggered by very low levels of exposure, including levels lower than permissible as established by the government for various chemicals and typically below the levels tolerated by most people. Symptoms are triggered by a wide range of substances found in the workplace and at home. Solvents, paint, varnishes, adhesives, pesticides, and cleaning solutions are most frequently implicated. Other substances include new building materials and furnishings, formaldehyde in new clothes, artificial fragrances in cleaning compounds and personal care products, detergents, car exhaust, and copying machine and laser printer toner. Symptoms can occur after inhaling chemical vapors, having chemicals touch the skin, or after ingestion. Sensitivity to a particular chemical can lead to sensitivity to an ever-widening range of other, often dissimilar, chemicals. This characteristic is known as the "spreading phenomenon."

It may be useful to think of MCS at one end of a spectrum that encompasses a wide range of chemical sensitivities. At one end are individuals who may suffer from mild symptoms, such as simple sinus congestion or headaches, which usually resolve when the triggering chemical agent is removed. At the other end of the spectrum are individuals with full-blown MCS who suffer extremely debilitating symptoms that can last for months or years after exposure.

Why do some people contract MCS and others, with the same level of exposure, do not? Because of biochemical individuality, all humans manifest disease according to individual genetic make-up, past chemical exposure, and overall general state of health which includes "total load." Total load refers to all stressors in a person's life, including chemical exposure, poor nutrition, emotional tension, allergies, infections, trauma, and physical stress.

Although the exact mechanism whereby chemicals create this phenomenon of heightened sensitivity has not yet been clearly elucidated, theories are emerging which will hopefully lead to greater understanding and better treatment of MCS.

Recent studies have demonstrated how toxins, having gained access to the brain by the olfactory nerve, can cause release of excitatory amino acids that result in swelling, dysregulation, and destruction of brain cells. The olfactory nerve is also the pathway to the limbic system which is an area of the brain where the nervous, immune, and endocrine systems interact. The limbic system regulates an extremely wide variety of body functions. Many of the varied and seemingly bizarre symptoms reported by persons with MCS are actually consistent with symptoms known in the medical literature to occur when various parts of the limbic system are damaged by either chemicals or physical injury.

Toxic chemicals can also cause direct damage to specific tissues of the body such as enzymes in the liver that are essential in the detoxification pathway. Because of inadequate amounts of detoxifying enzymes, the MCS person is less able to handle chemical loads. Next, recent data indicate that certain toxins in the environment, especially chlorinated compounds, mimic natural hormones, causing disruption of endocrine systems, such as the thyroid, adrenal, and reproductive systems.

The first documented cases of environmental illness resulted from widespread chemical poisoning during World War I. The exposure to mustard gas had long-term consequences for soldiers, many of whom developed chronic symptoms of chemical sensitivities. More recently, thousands of veterans who fought in the Gulf War returned with symptoms that were similar to those found in patients diagnosed with MCS.

Since World War II the production of synthetic chemicals has increased significantly. In 1945, the estimated worldwide production of these chemicals was fewer than 10 million tons. Today it is over 110 million tons. As more and more synthetic chemicals are introduced into the environment, larger numbers of healthy people are becoming affected. Most people with MCS have not been through a war. They have become ill from ordinary day to day, low level exposures from poor indoor air quality in their homes and workplaces. MCS sufferers often say that their role in society is like the canary in the coal mine. When the canary collapsed, the miners were warned that lethal gases were in the air.

Although MCS is a rapidly growing problem, sometimes called a silent epidemic, health care workers know little about the subject. Why? Chemical sensitivity is a relatively new field of medicine, controversial in nature, and not recognized or understood by most physicians. The illness does not fit neatly into the current medical model. And unlike diabetes or hypertension, there is no simple medical test that can be used to make the diagnosis. There are remarkably few individuals in medicine who have toxicology training and who are sensitive to the possible problems of neurological, behavioral, and psychiatric effects resulting from chemical exposures. In addition, the chemical and insurance industries have played a major role in influencing the average person's perceptions about chemicals and their impact on living organisms.

One of the most important aspects of treatment of the chemically sensitive person is to avoid or reduce toxic chemical exposures as much as possible in order to allow the body to heal. A healthy home is a prerequisite for those who wish to regain their health. The person with MCS needs a sanctuary of peace and well-being amid a world saturated with toxic chemicals.

In spite of widespread ignorance and vested financial interests, MCS is gradually becoming known to the public as more and more people are becoming ill. A small but growing number of physicians specializing in environmental medicine has been focusing on this serious problem for several years. If you would like information about a physician in your area with expertise in the diagnosis and treatment of chemically related health problems, call the American Academy of Environmental Medicine in New Hope, PA, at (215)862-4544. You can request a list of referrals as well as information.

Author Testimonials on MCS

Paula Baker

If someone had told me in the early years of my career that I would be writing a technical "how to" book about healthy homes, I would have looked at them with total incredulity! I would have explained that, as an architect, my main concerns were with the creation of beautiful and interactive spatial forms, that my aspirations were artistic rather than technical. It seems that fate had a different course for me. I joined the ranks of the chemically sensitive. Erica Elliott, my friend and physician, diagnosed my condition and helped me to get back on my feet. It was through her that I first heard of healthy building. She told me of the alarming number of chronically ill patients consulting with her who were diagnosed with MCS. For many, the primary cause of illness was exposure to multiple toxins in the home.

Even though I specialized in residential architecture, I had to admit I knew little about the health implications of standard home construction. After working with Erica to design her home, I began intensive research into this new frontier in architecture and building. It was then that my personal and professional life took a new direction.

main concerns were with the creation of beautiful and interactive spatial forms, that my aspirations were artistic not technical. It seems that fate had a different course for me. I joined the ranks of the chemically sensitive. Erica Elliott, my friend and physician, diagnosed my condition and helped me to get back on my feet. It was through her that I first heard of healthy building. She told me of the alarming number of chronically ill patients whom she was seeing with MCS. For many, the primary cause of illness was exposure to multiple toxins in the home.

Even though I specialized in residential architecture, I had to admit I knew little about the health implications of standard home construction. After working with Erica to design her home, I began intensive research into this new frontier in architecture and building. It was then that my personal and professional life took a new direction.

Once I learned the facts, I could never again allow certain products to be used in any projects with which I became involved. I understood the health threat that they posed to my clients, other inhabitants, construction workers, and the planet. As a result of my research, my goals as an architect have grown. In order to truly nurture us, our buildings must not only be beautiful in a spatial sense, they must also be healthful and conceived with mindfulness of limited planetary resources. The same building can destroy human health or enhance vitality. The difference lies in the materials and methods of construction.

Erica Elliott

My involvement with indoor air quality issues began in 1991 when I went to work for a large medical corporation as a family physician. The building which housed the clinic was new and tightly sealed, with nonoperable windows and wall-to-wall carpeting. Previously in excellent health, a world class mountaineer and marathon runner, I began to develop unexplainable fatigue. After several months, more symptoms developed which included rashes, burning eyes, chronic sore throat, and headaches. The symptoms subsided in the evenings after I left the workplace, only to return again upon reentering the building.

By the second year of employment, I had developed persistent migraine headaches, muscle and joint pains, insomnia, confusion, lack of coordination, memory loss, and mood swings. By then, the symptoms had become permanent, even when I was away from the workplace on weekends. My physician colleagues were puzzled by my symptoms. Some felt I was suffering an unusual manifestation of depression and would benefit from anti-depressant medication. These medicines were not helpful and only masked the problem. I finally had the good fortune of finding a physician trained in environmental medicine who believed that I had nervous and immune system damage related to chronic, low level exposure to poorly ventilated toxins in the workplace.

By the time I resigned my position on the staff of the clinic and the local hospital, I had a full-blown case of multiple chemical sensitivities, also known as environmental illness. Most synthetic chemicals commonly found in the modern world, even in minute amounts, caused me to have adverse reactions to such a degree that life became a painful ordeal. With diligent avoidance of toxins, abundant rest, detoxification therapies, and other measures, my life has begun to stabilize.

Since very few physicians are trained in toxicology and environmental medicine, I began immersing myself in this field of study, both to help myself as well as others. It wasn't long before my practice consisted primarily of patients with immune dysfunction, including multiple chemical sensitivity, auto-immune disease, chronic fatigue syndrome, fibromyalgia, and severe allergies. I was struck by the number of patients who dated the onset of their symptoms to a move to a new home or to the remodeling of a school or office. They invariably had been to many doctors who treated them for conditions such as allergies, asthma, sinusitis, and depression. The underlying causes were not identified. By the time the correct diagnosis was made, the patients' immune systems were often severely, sometimes irreversibly, damaged.

It was during my own recovery that I decided to build a home using nontoxic building materials. Paula Baker was my architect, patient, and neighbor. Together we began researching various available products and associated health effects. Shortly thereafter, we had the pleasure of meeting John Banta, the beginning of a fruitful collaboration.

John Banta

My introduction to the downside of indoor air quality occurred along with my introduction to fatherhood in 1980. Like many first-time parents, my wife and I wanted to welcome our newborn by decorating the nursery. We painted and carpeted the room in anticipation of our new arrival. The room smelled of chemicals; I noticed that I did not feel good in there. But it wasn't until our baby became ill that I realized what a serious problem we had created. By the time I made the connection between my daughter's medical condition and the toxins in the nursery, she had become sensitized to even minute amounts of toxic chemicals commonly found in the environment and was in severe distress. My wife and I decided to buy an old Victorian home that had not been remodeled in over 40 years. We proceeded to convert the building into a chemical-free sanctuary where our daughter could begin to heal from her devastating illness.

During that time I was working as a medical technician in a research lab where I was exposed to numerous toxic chemicals including formaldehyde, benzene, toluene, xylene, and several disinfectants. Over the next four years, I felt progressively worse while at work, yet I would feel better once I returned to our carefully remodeled home. My job related health problems finally became so severe that I made the difficult decision to quit. Little did I know that a new and exciting career was awaiting me.

Because of my hands-on experience in renovating my own healthy home, people began to ask me for my advice. My wife urged me to begin consulting professionally which I have done full-time since 1986.

Thousands of people have consulted me over the years about their homes. Typically, I am contacted in the middle of a disaster: "the walls are moldy," "the paint is causing headaches," or "the landlord sprayed pesticides to control insects." I am then hired to determine the cause and suggest a remedy for the problem. My job often includes educating a skeptical landlord or spouse as to the causal relationship between the health of the occupant and the problem in the home.

The most rewarding work for me is consulting during the planning phase of new construction, where I can help my clients prevent problems before they occur. Although I do not design or build homes, I can troubleshoot and monitor to help ensure a nontoxic, healthful, and nurturing abode.

I have really enjoyed working with Paula and Erica in creating this book. For me, it offers a way to reach more people by providing them with the information they need to create a healthy home from the outset.

Appendix B. Page References to Brand Name Products and Services

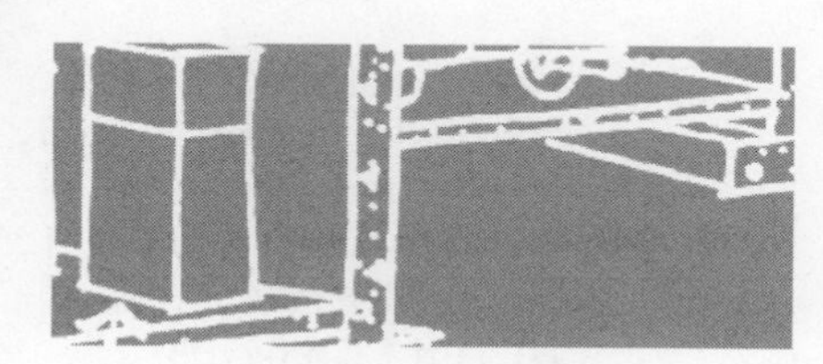

Appendix C. Manufacturers, Service Providers, and Catalog Sources

Manufacturers, service providers, and catalog sources for products and services recommended throughout the book are alphabetically listed below.

Company name	Telephone	Products/services	References
AirChek Inc.	(800)247-2435	Home test kits for radon, lead, formaldehyde, water purity, and microwave leakage	Divisions 7, 13
Allergy Relief Shop	(800)626-2810	Catalog sales of supplies for allergy free home	General Information, Divisions 12, 13
Allergy Resources	(800)873-3529, (719)488-3630	Catalog sales of cleaning compounds and bodycare products	General Information
Allermed Corporation	(972)442-4898	Air filters	Division 15
Alpha Labs	(800)769-3754	Gaussmeter	Division 16
Alsto's Handy Helpers	(800)447-0048	Mail order house	Division 7
American Environmental Health Foundation	(800)42802343, (214)361-9515	Catalog sales of household, building, personal care, and medical products, as well as clothing, books, and vitamins	General Information

Company name	Telephone	Products/services	References
American Formulating and Manufacturing	(800)239-0321, (619)239-0321	Wood finishes; adhesives, paints, sealers, and primers for diverse surfaces; carpet shampoo and sealer; vapor and waterproof barrier; caulking compound; paver seal; concrete waterproofing; cleanser/degreaser; mildew control; disinfectants	Divisions 1, 3, 6, 7, 8, 9
Apthecure, Inc.	(800)969-6601	Spectracidal disinfectant	Division 1
Aqua Zone	(847)439-4454	Water ozonation system	Division 13
Architectural Forest Enterprises	(800)483-6337, (415)822-7300	Hardwood plywood veneers	Division 6
Ashland Chemical Inc.	(800)258-0711, (614)889-3333	Sealer for concrete floors	Divisions 3, 9
Aubreys Organics	(800)282-7394	Catalog sales of bodycare products derived from herbs and vitamins	General Information
Bangor Cork Company	(610)863-9041	Natural cork flooring	Division 9
Benjamin Moore & Company	(800)344-0400, (201)573-9600	Latex paint	Division 9
Borden Inc.	(800)426-7336, (800)848-9400	Glue	Divisions 6, 8
Bright Futures Futons	(888)645-4452, (505)268-9738	Futons and couch beds	Division 12
BRK Electronics	(800)392-1395, (630)851-7330	Smoke detector	Division 16
Brookstone Company	(800)846-3000	Mail order house	Division 7
Building for Health Materials Center	(800)292-4838	Catalog sales of building products	General Information
C-Cure Corporation	(800)895-2874, (713)492-5100	Tile setting, dry-set mortar, latex-cement mortar, dry tile grout	Division 9
Ceramic Radiant Heater Corporation	(800)331-6408	Ceramic radiant heater	Division 15
Cervitor Kitchens Inc.	(800)523-2666, (818)443-0184	Metal kitchen cabinetry	Division 6
Cetco	(800)948-5419, (708)392-5800	Wall panels	Division 7
Chem Safe	(210)657-5321	Paints mixed to order	Division 9
Chemical Specialties Inc.	(800)421-8661	Pressure treatment for wood	Division 6
Coronado Paint Company	(800)88304193, (904)428-6461	Latex paint	Division 9
The Cotton Place	(800)444-2383, (714)494-3002	Fabrics	Division 12
Crate and Barrel	(800)323-5461	Solid wood, glass, and metal furnishings and accessories	Division 12
CrossLink	(888)301-9663, (505)983-6877	Wood products	Divisions 6, 9

Company name	Telephone	Products/services	References
The Cutting Edge Catalog	(800)497-9516	Sells pesticide checking kits	Division 10, 13
DAP/Dow Corning Products	(800)634-8382	Sealant	Division 7
Davis Colors	(800)356-4848, (213)269-7311	Pigments for concrete	Division 3
Delta Products Incorporated	(617)471-7447	Caulk sealing cord	Division 7
Denny Sales Corporation	(800)327-6616, (954)971-3100	Foil vapor barrier	Division 7
Dr. Sherry Rogers	(315)488-2856	Mold and formaldehyde test kits.	Division 13
E.L. Foust Company, Inc.	(800)225-9549	Air filters	Division 15
Eco Design/Natural Choice	(800)621-2591, (505)438-3448	Oil primer and finish for interior hardwoods; finishes, sealers, undercoats, and primers for wood, stone, terra cotta, and linoleum; cork flooring, elastic glue, pigment powders, thinner oil for priming woods; wood preservative; cleaning agents, smoke detector. Natural Choice is a mail order catalog distributed by Eco Design.	Divisions 1, 6, 8, 9, 16
Eco Products Inc.	(303)449-1876	Building product supplier	General Information
Eco Timber International	(510)549-3000	Sustainably harvested wood source	Division 6
Ecological Beginnings	(440)543-3180	Solid hardware nursing furnishings	Division 12
Ecoworks	(800)466-9320, (410)448-3317	Smoke detectors	Division 16
Elastomerics Corporation	(800)621-7663, (413)533-8100	Roofing membrane	Division 7
Environmental Health Center	(214)373-5149	Home test kits for mold	Division 13
Environmental Testing & Technology	(760)436-5990	EMF measurements and mitigation consultation	Division 16
Environmentality World	(800)288-0230	Water purification systems for pools and spas	Division 13
Eternit, Inc.	(800)233-3155, (215)926-0100	Backer board	Division 9
Faultless Starch/Bon Ami Company	Wide distribution	Cleaning agents	Division 1
Fiber-Lock Company	(800)852-8889, (817)498-0042	Fiber additive reinforcement for concrete slabs	Division 3
Forbo Industries	(800)233-0475, (717)459-0771	Linoleum and cork flooring	Division 9
Forest Trust Lumber Brokerage	(505)983-3111	Lumber broker	Division 6

Company name	Telephone	Products/services	References
Franklin International	(800)347-4583	Adhesives	Divisions 6, 8
Furnature Inc.	(617)782-3169	Mattresses, upholstered sofas and chairs	Division 12
GE	(800)255-8886	Sealant	Division 7
Gerbert Ltd.	(800)828-9461	Linoleum	Division 9
Glidden Company	(800)221-4100, (216)892-2900	Paint	Division 9
Glouster Company	(800)343-4963, (508)528-2200	Caulks and sealants	Division 7
The Good Water Company	(800)471-9036, (505)471-9036	Water filtration systems and consultation	Division 11
Greenwood Cotton Industries	(800)546-1332	Batt insulation	Division 7
Hague Quality Water International	(614)836-2195	Whole house water purification system	Division 11
Heart of Vermont	(800)639-4123	Bedding and other products for the chemically sensitive	Division 12
Hendricksen Naturlich	(800)329-9398, (707)824-0914	Natural fiber carpeting, underpads, adhesives	Division 9
Home Improvements	(800)642-2111	Mail order house	Division 7
Homespun Fabrics and Draperies	(800)251-0858	Fabrics	Division 12
Huntar Company Inc.	(650)873-8282	Guassmeters	Division 16
Hydrocote Company Inc.	(800)229-4937, (908)257-4344	Wood stains and polyurethane clear coat	Division 9
Icynene Inc.	(800)946-7325	Spray-on foam insulation	Division 7
Innovative Formulations	(800)346-7265, (520)628-1553	Paints, fluid applied roofing system	Divisions 7, 9
International Cellulose Corporation	(800)444-1252; (713)433-6701	Spray in or loose fill insulation	Division 7
International Grating Inc.	(800)231-0115, (713)633-8614	Fiberglass rebar	Division 3
International Med Com	(808)823-0336	Device for measuring radioactivity.	Division 13
James Hardie Company	(800)426-4051, (714)356-6300	Tile backer board	Division 9
Janice Corporation	(800)526-4237, (201)691-2979	Mattresses and bedding	Division 12
Kendall/Polyken	(800)248-0147, (508)261-6200	Aluminum tape air barrier	Division 7
L.M. Scofield Company	(800)222-4100	Pigments for concrete	Division 3
Leggett & Platt	(800)426-4051, (714)356-6300	Jute underpadding	Division 9

Company name	Telephone	Products/services	References
The Living Source	(800)662-8787, (817)776-4878	Catalog sales of products for chemically sensitive and environmentally conscious	General Information, Division 9
LM Scofield Company	(800)800-9900, (213)723-5285	Building caulk	Division 7
Louisiana Pacific	(800)579-8401	Gypsum sheating	Division 6
Magnetic Sciences International	(800)749-9873, (520)822-2355	Gaussmeter	Division 16
Manufactured Plastics and Distribution Inc.	(303)296-3516, (954)971-3100	Air barrier, radon barrier	Division 7
Mapei Inc.	(800)621-6491, (241)271-9500	Joint grout	Division 9
Marshall Vega Corporation	(501)448-3111	Non-metallic rebar	Division 3
Medite Corporation	(800)676-3339, (503)773-2522	Fiberboard	Divisions 6, 9
Miller Paint Company	(800)852-3254	Paint	Division 9
Mold Survey Service	(315)488-2856	Mold testing	Division 13
Murco Wall Products	(800)446-7124, (817)626-1987	Joint cement, wall paint	Division 9
The Natural Alternative	(612)351-7165	Upholstered furniture	Division 12
The Natural Bedroom	(800)365-6563, (415)920-0790	Bedding, bedroom furniture	Division 12
The Natural Choice	(800)621-2591	Catalog sales paints, stains, and other home products	General Information
Natural Cork C°	(800)404-2675, (404)872-4168	Cork flooring	Division 9
NEEDS	(800)634-1380	Catalog sales of personal care products for chemically sensitive	General Information
Neff Kitchen Manufacturers	(800)268-4527, (905)791-7770	Manufactured cabinets	Division 6
Nigra Enterprises	(818)889-6877	Broker for air and water filtration systems, heaters, vacuum cleaners	Divisions 11, 15
Nisus Corporation	(800)789-4248, (603)659-5919	Wood preservative	Division 6
Nontoxic Environments Inc.	(800)789-4348, (603)659-5919	Catalog sales of building, household and personal products for the chemically sensitive and earthwise	General Information, Division 9
The Non-Toxic Hot Line	(800)968-9355 (orders only), (510)472-8868	Catalog sales of products oriented to indoor air quality and safety for homes, offices, and cars	General Information, Division 1
Nordic Builders	(602)892-0603	Foam insulation	Division 7

Company name	Telephone	Products/services	References
Okon Inc.	(800)237-0565, (303)232-3571	Clear sealer, wood stain and water repellant	Divisions 6, 8, 9
Ostermann & Scheiwe	(800)344-9663	Wood stains, preservative	Divisions 6, 8, 9
Owens Corning	(800)438-7465	Fiberglass insulation material	Division 7
Ozark Environmental Services	(800)835-8908	Air and water testing; consultation on toxic gases, molds, asbestos, VOCs, pesticides, gas leaks, EMFs and radon	Divisions 13
Pace Chem Industries	(800)350-2912, (805)686-0745	Seal for hardwood floors	Division 9
Palmer Bedding	(214)335-0400	Bedding products	Division 12
Palmer Industries Inc.	(888)685-7244, (301)898-7848	Sealer and primer for gypsum board; wood preservative	Divisions 6, 9
Panolam	(888)726-6526, (541)928-1942	Melamine board	Division 6
Pella Corporation	(800)547-3552, (515)628-6457	Windows	Division 12
Perma-Chink Systems	(800)548-3554	Wood preservative	Division 6
Planetary Solutions	(303)442-6228	Building materials supplier	General Information
Plaza Hardwood, Inc.	(505)466-7885	Sustainably harvested and recycled wood	Division 6
Positive Energy	(800)488-4340, (303)444-4340	Mastic	Division 15
Pottery Barn	(800)922-5507	Solid wood furniture, glass and metal furniture and accessories, cotton window dressings	Division 12
Professional Discount Supply	(800)688-5776, (719)4440-0646	Soil gas collector matting	Division 7
Quality Wood Products	(505)756-2744	Sustainably harvested wood	Division 6
Radiation Technology	(216)666-7710	Measurement devices for electric and magnetic fields	Division 16
Rainforest Alliance	(888)MY-EARTH, (212)677-1900	Nonprofit organization setting standards for sustainable wood harvesting. Provides list of sources.	Division 6
Real Goods	(800)762-7325	Smoke alarms; water purification system for pools	Divisions 13, 16
ReCon Restoration Consultants	(916)736-1100	EMF detection and remediation	Division 16
Resource Conservation Technologies, Inc.	(410)366-1146	Roll-on paint roofing	Division 7
Roy Akers Advanced Foil Systems	(800)421-5947, (909)390-5125	Air barrier	Division 7

Company name	Telephone	Products/services	References
Rubber Polymer Corporation	(800)860-7721	Concrete waterproofing membrane	Division 3
Santa Fe Heritage Door Company	(505)983-5986	Custom wood doors	Division 8
Seventh Generation	(800)456-1177	Bedding, shower curtains	Division 12
Shelter Supply	(800)762-8399, (612)854-4266	Ventilation products	Division 15
Sherwin Williams Company	(800)321-8194, (216)566-2902	Interior latex paint	Division 9
Siemens	(800)964-4114	Electrical panels and subpanels	Division 16
Sinan Company	(916)753-3104	Exterior wood finish, shellacs, indoor/outdoor enamels, linoleum flooring adhesive, binder	Divisions 8, 9
Skanvahr Coatings Ltd.	(800)329-3405, (509)466-1841	Wood flooring finishes	Division 9
Solutions	(800)342-9988	Mail order, stone surface finish	Division 9
Spanish Pueblo Doors	(505)473-0464	Custom wood doors and cabinets	Division 8
Spectra-Tone Paint Corporation	(800)272-4687, (909)478-3485	Latex paint	Division 9
Star Bronztec	(800)321-9870	Woodwork finish	Division 9
Statements	(505)988-4440	Jute pads	Division 9
Stocote Products Inc.	(800)435-2621, (815)675-6713	Air barrier	Division 9
Stocote Products Inc.	(800)435-2621, (815)675-6713	Barrier sheeting and sealers	Division 7
Streit's Healthy House Service	(505)898-8432	EMF assessment and solutions; remote circuit shut-off switch	Divisions 13, 16
Synthetic Industries	(800)621-0444, (423)899-0444	Fiberglass reinforcement for concrete slabs	Division 3
Thoro Systems Products	(800)322-7825, (904)828-4900	Concrete surface waterproofing	Division 7
U.S. Borax Corporation	(800)553-4872, (310)522-5300	Wood preservative	Division 6
U.S. Intec Inc.	(800)331-5228, (800)624-6832	Elastomeric roof coating, applicator source for modified bitumen roofing	Division 7
United Gilsonite Laboratories	(800)845-5227, (717)344-1202	Polyurethane flooring finish	Division 9
W.R. Meadows	(800)342-5976, (708)683-4500	Concrete sealer	Division 3
Watercheck, National Testing Laboratories, Inc.	(800)438-3330	Water testing	Division 11
Weldwood of Canada Inc.	(888)566-4522, (905)542-2700	Hardwood veneered plywood panels	Division 6

Company name	Telephone	Products/services	References
WF Taylor Company Inc.	(800)397-4583, (909)360-6677	Adhesives, mastics, and seaming tapes	Divisions 8, 9
Wirsbo	(800)321-4739	Polyethelene tubing for radiant floor heating; polyethelene plumbing system	Division 15
Wm. Zinsser & Company	(732)469-8100	Primer sealer	Division 9
Woodworkers Supply	(800)645-9797, (505)821-0500	Catalog sales of building materials	General Information, Divisions 6, 9
Xypex Chemical Corporation	(800)961-4477, (604)273-5265	Concrete waterproofing	Division 3

INDEX

D

E

H

I

N

O

S

InWord Press Distribution

End Users/User Groups/Corporate Sales

InWord Press books are available worldwide to end users, user groups, and corporate accounts from local booksellers or from SoftStore Inc. Call toll-free 1-888-SoftStore (1-888-763-8786) or 505-474-5120; fax 505-474-5020; write to SoftStore, Inc., 2530 Camino Entrada, Santa Fe, New Mexico 87505-4835, USA, or e-mail orders@hmp.com. SoftStore, Inc., is a High Mountain Press company.

Wholesale, Including Overseas Distribution

High Mountain Press distributes InWord Press books internationally. For terms call 1-800-466-9673 or 505-474-5130; fax to 505-474-5030; e-mail to orders@hmp.com; or write to High Mountain Press, 2530 Camino Entrada, Santa Fe, NM 87505-4835, USA.

Comments and Corrections

Your comments can help us make better products. If you find an error, or have a comment or a query for the authors, please write to us at the address below or call us at 1-800-223-6397.

InWord Press, 2530 Camino Entrada, Santa Fe, NM 87505-4835 USA

On the Internet: http://www.hmp.com